FINANCING SCHOOLS AND EDUCATIONAL PROGRAMS

D0144281

Other Titles of Interest

Financial Accounting for School Administrators, by Ronald E. Everett, Donald R. Johnson and Bernard W. Madden, 978-1-61048-771-9

Cutting Costs and Generating Revenues in Education, Second Edition, by Tim L. Adsit and George R. Murdock, 978-1-60709-897-3

Planning Educational Facilities: What Educators Need to Know, Third Edition, by Glen I. Earthman, 978-1-60709-046-5

Equalize Student Achievement: Prioritizing Money and Power, by Ovid K. Wong and Daniel M. Casing, 978-1-60709-145-5

FINANCING SCHOOLS AND EDUCATIONAL PROGRAMS

Policy, Politics and Practice

AL RAMIREZ

ROWMAN & LITTLEFIELD EDUCATION
A Division of
ROWMAN & LITTLEFIELD PUBLISHERS, INC.
Lanham • New York • Toronto • Plymouth, UK

Published by Rowman & Littlefield Education
A division of Rowman & Littlefield Publishers, Inc.
A wholly owned subsidiary of The Rowman & Littlefield Publishing Group, Inc.
4501 Forbes Boulevard, Suite 200, Lanham, Maryland 20706
www.rowman.com

10 Thornbury Road, Plymouth PL6 7PP, United Kingdom

Copyright © 2013 by Al Ramirez

All rights reserved. No part of this book may be reproduced in any form or by any electronic or mechanical means, including information storage and retrieval systems, without written permission from the publisher, except by a reviewer who may quote passages in a review.

British Library Cataloguing in Publication Information Available

Library of Congress Cataloging-in-Publication Data

Ramirez, Al, 1947–
  Financing schools and educational programs : policy, politics, and practice / Al Ramirez.
    p. cm.
  Includes bibliographical references and index.
  ISBN 978-1-4758-0176-7 (cloth : alk. paper)—ISBN 978-1-4758-0177-4 (pbk. : alk. paper)—ISBN 978-1-4758-0178-1 (electronic) 1. Public schools—United States—Finance.
2. Government aid to education—United States. I. Title.
  LB2825.R33 2012
  379.1'210973—dc23   2012026081

♾™ The paper used in this publication meets the minimum requirements of American National Standard for Information Sciences—Permanence of Paper for Printed Library Materials, ANSI/NISO Z39.48-1992.

Printed in the United States of America

To Adrienne and Erika and all the other
wonderful teachers who succeed with reluctant learners.

TABLE OF CONTENTS

Preface

PROFESSORS AND INSTRUCTORS WHO TAKE on the challenge of teaching a graduate course on school finance know how difficult it is to find an appropriate book for their course. Part of the reason it is so difficult is that the states and territories each have their own method of collecting revenue and distributing resources to school districts. In addition, school finance scholars and practitioners understand that education finance is more than numbers and spreadsheets; it involves policy, politics and professional practice.

This book is written with the idea that the individual who teaches the course is an expert in the field and can guide students in this understanding. The book is designed as a resource for the instructor to facilitate overall course goals and objectives. It is expected that the professor who leads the course will bridge the variance between the subject matter presented in the book and the unique situation in his or her state and region.

This book is written on the subject of how the public schools in the United States are financed and how other funds are raised for educational programs in elementary and secondary schools. While there is a logical sequence to the chapters, it is not unexpected that instructors will choose to assign chapters in a sequence that matches their course syllabus. The book spans both the theoretical and practical aspects of the topics presented. The text is written primarily for graduate students in programs for education administration, policy studies, public administration, public finance and public accounting. Each chapter is structured so as to enhance the book's value to pre-service students preparing for entry-level school administration positions as well as candidates for advanced degrees who need more research-based theoretical content. The book can also serve as a resource for practitioners and education policy leaders, e.g., school board members, foundation program officers and legislators at the local, state and national levels.

Outstanding Features

Supplemental materials to support the book are available online to professors who adopt the book. These materials, presented in a digitized format, include: PowerPoint presentations for each chapter; chapter lesson outlines; test item banks for each chapter; recommended chapter and term projects; problem-based learning projects (PBLs); and related projects, e.g., simulated school-based budget, enrollment projection, grant-writing exercise or calculating local school taxes.

These exercises include class activities for individuals and small groups. Appendices and supplemental materials include budget forms, sample materials and student project assignments. Selected charts, graphs and tables are included as appropriate. A section on adapting the text and materials for use as an online or blended course is also provided. In essence, the text and supplemental instructor materials can function as a turnkey course on school finance.

The URLs listed here can serve as a glossary for terms included in this textbook: http://nces.ed.gov/programs/digest/d10/app_b.asp; http://nces.ed.gov/pubs2012/2012313/appendix_b.asp; http://nces.ed.gov/pubs2009/fin_acct/index.asp.

Introduction

THE TOPIC OF EDUCATION FINANCE OFTEN EVOKES images of spreadsheets, accounts and endless columns of numbers. But in fact, the topic involves much more. This is a book on the subject of how the public schools are financed and how supplementary funds are raised for educational programs in public elementary and secondary schools. The subject matter is presented in a way that considers the policy origins of funding programs, while addressing the politics associated with the policy formation. In addition, an effort is made to explain the practical application of these funding policies by offering information about how these policies function in practice.

The book spans both the theoretical and practical aspects of the topic. The text is written primarily for graduate students in programs for education administration, policy studies, public administration, public finance and public accounting, although those interested in the topics of education funding and policy will find it of value as well. Each chapter is structured so as to enhance the book's value to pre-service students preparing for entry-level school administration positions as well as candidates for advanced degrees who need more research-based theoretical content. The book can also serve as a resource for practitioners and education policy leaders, e.g., school board members, foundation program officers and legislators at the local, state and national levels.

Here is a brief summary of the book's contents:

Chapters 1, 2 and 3 present the big themes considered in the book. The chapters strive to set a conceptual framework for the material studied in the book by laying out the broad policy issues related to school finance, offering a historical context to help explain how things got to be as they are today, and by addressing directly the fundamental question about whether money matters in education.

Chapters 4, 5, 6 and 7 consider how scholars, policy makers and the courts have defined, circumscribed and interpreted the major school finance issues over time. These chapters provide a grounding in the key aspects of school funding.

Chapters 8 through 13 consider the practical aspects of collecting, allocating and accounting for the hundreds of billions of dollars that flow through the PK–12 system each year. These chapters reveal the "nuts and bolts" of funding schools across the country. The chapters look at funding from top to bottom, that is, from the federal and state level to the classroom.

Chapter 14 speculates about the future of school finance in light of the topics and issues considered in the book. The chapter contemplates policy implications and the potential impact on the policy maker and the practitioner school administrator alike. Commentary about future policy issues, political maneuvering and trends in practice are proffered.

Financing Schools and Educational Programs: Policy, Politics and Practice is a comprehensive work designed to provoke thought about key school finance policy issues while clarifying the often arcane and opaque dimensions of the topic. Insights into the world of policy, politics and practice are presented within the context of money for schools. The goal of the work is to make the nuances of school finance accessible to a wide audience of policy makers, aspiring scholars and education practitioners.

Dimensions of Education Finance **1**

Aim of the Chapter

THIS CHAPTER PRESENTS A CONCEPTUAL FRAMEWORK of school finance designed to facilitate a deeper understanding of the essential issues and knowledge associated with the topic. The chapter also lays out a theoretical foundation, which provides the nexus for the framework and is part of the multidimensional topics presented throughout the book. The material in this chapter provides a context and serves as a pre-organizer in anticipation of the topics, terms, theories and concepts covered in subsequent chapters. It challenges the reader to think about the big questions in school finance.

Introduction

When those new to the study of school finance begin learning about the subject it is common for them to expect the topic to be devoted solely to budgets and accounting. Novice policy makers, new school leaders and many graduate students are often surprised to find that a discussion of education funding often covers a wide-ranging collection of topics. Clearly, the consideration of figures and formulas is important to the field of education finance, but understanding the context for those numbers is equally important to those who would shape policy or lead an education organization.

Many factors come into play before the final figures are set for an annual budget, be it at the federal, state, school district or school level. School budgets don't exist in a vacuum, devoid of outside influence. Nor are they based exclusively on need or even the fair distribution of resources. Many of those who are involved in pre-kindergarten through high school education (PK–12) and work with children on a daily basis commonly assert that the children in the schools are the priority: "we are here for the kids." They are puzzled when those outside the classroom talk about budget cuts, competing priorities, market systems, privatization or efficiencies. Conversely, political leaders, policy makers and government workers remote from the schools are sometimes uninformed about school needs and the best uses of available resources to support higher levels of learning in the schools. What is clear in all this is that there are many perspectives around the questions of funding for schools.

As a nation, Americans have decided that the idea of "the common school" was one worthy of public support through taxation. The public school system of today, which has educated 90 percent of the U.S. population for the past century, did not spring up whole along with Christopher Columbus or the Liberty Bell. Nearly a two-century-long struggle to establish tuition-free, tax-supported, pre-collegiate education preceded the public school system seen today in every village, town and city across America. School finance, it seems, has a historical dimension as well. Remarkably, those familiar with the historical development of the public schools in the nation marvel at the cyclical nature of the issues and policy questions that confront school leaders and policy makers today.

But history has also delivered to Americans a bountiful legacy in the form of a set of national values, manifested in the U.S. Constitution, that also serve to shape modern-day issues of school finance. As a people, Americans have a strong sense of fairness. We see this in such cultural artifacts as our love of competitive sports or in our jurisprudence system. Today, we find ourselves in a century-long dialogue about equity in school finance, equal educational opportunity and the adequacy of resources to meet our ideals. Fairness is a core American value that permeates the civic discourse about money for schools in contemporary society. This discourse uses terms like "equity" and "adequacy" to frame the debate, but at the heart of the matter, we are struggling to decide what is fair.

Theoretical and philosophical questions have their place in this dialogue, but money has a way of inserting a large measure of practicality into debates about funding for schools. Thus, when questions of taxation, revenues and values intersect, politics inevitably moves to the center stage (McDonnell, 2009). School finance often boils down to the question of "who gets what," which remains the essential issue in so many state legislative sessions at the close of these annual convocations. The public schools and their funding needs are the big-ticket item in the states and territories of the nation. Policy leaders in this political environment are naturally preoccupied with questions of school finance.

Within our tripartite system of government the courts are the ultimate arbiters of fairness and legality in society. As such, the courts have played a major role with regard to questions of funding equity for schools. They have defined concepts like equal educational opportunity and crafted legal doctrines that guide the funding systems in the states. Over time a legal framework has been built, which informs policy development for the education system and substantially influences questions about funding for schools.

But one factor that defies human control is the economy. The cycles of boom and bust that plague both global and local economies also affect funding for schools.

So despite the best efforts of policy makers and school leaders to craft programs, plan budgets and design a system of schooling to serve the nation, economic realities force decisions about money for schools toward fundamental questions about priorities and available resources. The "Great Recession" of 2008–09 is still impacting the resource base for public education in most states across the nation.

All of these issues and many more interact in a dynamic environment, which ultimately shapes the practices associated with school finance. We see these influences in how money is collected, distributed and used for the support of the system of education that exists in the nation. When, where, how and for whom money is spent for education are questions that derive from this dynamic environment made up of policy, politics, history, jurisprudence, economics and practice.

Policy

The term "policy" is an elusive concept that is often misunderstood and misapplied. Part of the reason why policy is so amorphous is that it means different things to different people. Even policy makers differ in their definition and understanding of policy. Thus, a central question to our study of school finance is—what is policy?

Appealing to the dictionary helps some in our understanding of policy. The *Oxford English Dictionary* (2009) defines policy several ways. Here are two ways that seem more useful: "*A principle or course of action adopted or proposed as desirable,*

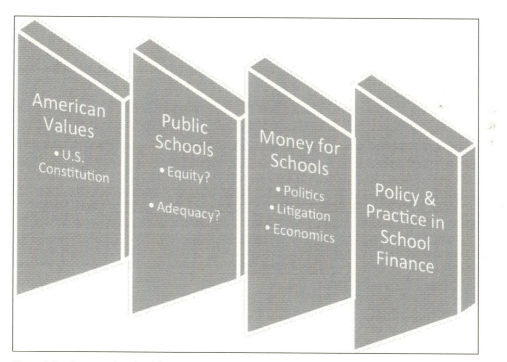

Figure 1.1   Concepts in school finance.

advantageous, or expedient; esp. one formally advocated by a government, political party, etc. Also as a mass noun: method of acting on matters of principle, settled practice" (para 4). As can be seen, the term is a bit slippery. It can be adopted or proposed, two very different states. It can be about principle or practice. And while generally associated with government, it can come from other sources—for example, a parent's "policy" about their adolescent child's access to the family car.

Within the realm of school finance, policy is seen in multiple aspects as well. It is derived from many sources: federal, state, local, school, the courts, custom and past practice. It is stated and sometimes explicit, as sometimes seen in statute or regulation. But finance policy can also be broad and nebulous—for example: our national commitment to an "equal educational opportunity" for all; state constitutional language regarding a "thorough and uniform system;" even the "free public schools" has various meanings, i.e., what is free?

One consequence of the often unstructured nature of policy is that it lends itself to controversy and debate (Stone, 1997). Within the realm of education finance one finds that big questions revolve around policy issues such as: equity in the distribution of resources; adequacy in the funding of education; what educational programs should be funded; who should benefit; what is the state's financial obligation; what financial responsibility belongs to the family; and where are the boundaries between the public good and private good of state-funded education?

These questions are but a part of the dialogue about school finance policy. Many stakeholders from many levels of society participate in the discussion. The consequences of how such policy questions are answered affect individuals and the nation as a whole. They can have an immediate effect or determine the course of events for generations. Policy is part of the either-or atmosphere within which education finance exists.

Politics

Education is a magnet for politics and political controversy. The reason for this is that education brings together the two issues most likely to start a fight—money and values (Wirt and Kirst, 2001). From the school finance perspective the political turf is clearly defined. The turf is bounded by questions like: what will the government provide as its education system; how will it be supported; who will pay to support it; and who will benefit from the system of education? Politics is the means of sorting out such questions.

Today we have a pre-collegiate system of schooling that spends over $650 billion annually. Almost forty-nine million students benefit from the education provided, millions are employed in the schools and countless others work in industries that serve the education system (National Center for Education Statistics [NCES], 2010). Manufacturers who build everything from buses to pencils have an interest in the financial health of the schools. Investment bankers on Wall Street handle the trillions of dollars set aside for teacher pensions. Local construction companies

Picture 1.1   Fights over money and values are the essence of politics.

are interested in the vote for the latest bond election in the local school district to build new schools.

Education is big business, very big business. In many communities across the nation schools are the biggest employer, have the biggest payroll, run the largest transportation system, serve more meals than any restaurant and collect more taxes than any other local government entity. Not surprisingly, people are interested in who will benefit, or profit, from this large enterprise.

Furthermore, the schools have a large stakeholder group, which comprises everyone in the community, state and nation. Modern society has come to recognize the essential function of a PK–12 system of schools to the viability of the nation. Pre-collegiate education is the foundation for the nation's human capital development. No country can expect to have a modern economy or a functioning civil society without a quality universal elementary and secondary system of schools (Checchi, 2008). Because of this, interest in the schools is seen from all quarters.

It should be expected, then, that different values will emerge regarding the role and mission of the schools, who they should serve and how they should be financed. On the one extreme are those who assert that education is a private good that should not be paid for through taxes, but rather each family should be responsible for the education of its children. On the other extreme are advocates who claim that only a public education should exist, to the exclusion of any private option, and this should be totally financed with public money. Add to this mix issues about what should be taught and who should teach it, and a smoldering political cauldron quickly emerges.

The U.S. Constitution

The forebears of the American republic laid a solid foundation for the new nation that is deeply rooted in individual liberty. When they designed the U.S. Constitution a chief aim of their effort was to protect citizens against an overbearing government and guarantee their freedom. We see this in the first ten amendments to the Constitution, known as the Bill of Rights. It was an essential proposal prerequisite to gaining ratification of the Constitution by the member states. This concept of liberty and individual rights is seen in the historical development of the system of education today in the United States, the means by which it is financed and the benefits derived by the people.

The founders understood that the bold experiment in self-governance hinged on the critical issue of whether a people could exert the prudence and restraint needed to make wise choices for their communities and the nation as a whole. It was understood that a democratic form of government required the participation of individuals who were sufficiently educated to deal with the processes of civic engagement.

Yet, no provision was made for a national system of schools and no mention of education was included in the Constitution then or since. This irony is explained by a key component of the governance structure of the nation, which divides powers and responsibilities among the various branches and levels of government, the states and the citizens themselves.

The states, communities and individual families were left to devise the methods and means of preparing the next generation for the duties of citizenship. Early on many state and local governments chose to create a role for themselves in the education of their children and youth. As this nascent education system grew, so did a body of law related to citizen, taxpayer, student, parent and teacher rights.

Over time a legal framework of school law materialized, which has shaped the system of education during its development and as it exists today. Questions about who would be educated, what would be taught and how it would be funded have all been vetted against the larger question of their constitutionality. The search for balance between the needs of society and the rights of the individual within the realm of education continues to this day. The U.S. Constitution has served as the guiding light in this search.

History

An important part of the curriculum at the U.S. Military Academy, West Point, is history. Military commanders know that there are many valuable lessons to be learned from past conflicts and from past and ancient civilizations. Within the military arena important strategies and tactics learned from history are incorporated into modern fighting policies. Military tradition is rooted in history and its special significance often has a practical reason found somewhere in the past. Would-be education and policy leaders are well advised to have a deep understanding of history and particularly the history of education. The lessons of history are frequently good sources of information when designing policy and carrying out policy (Neustadt and May, 1986).

Not only does an understanding of the past provide valuable knowledge about what has been tried before, but it can also lead us to what needs to be avoided. Furthermore, history makes available a context for understanding why things are the way they are. This is especially important in education and the financing of schools. Public education is susceptible to fads; in fact, it is not an overstatement to say it is plagued by fads. Education suffers from the "good idea" syndrome by which one "good idea" is layered on top of another, often without regard to context or a deep understanding of implications and unintended consequences. As a result, much time is wasted and resources squandered.

For example, the charter school movement looks like a sweeping innovation. However, many policy makers are surprised to learn that local, state and federal government entities have been chartering schools since the colonial era in America. How many education policy leaders can explain why Nebraska has 271 school districts, while Florida, with ten times the student population, has 67 school districts (NCES, 2010)? When did teacher certification come into existence and why was it established? What are the origins of the National School Lunch Program and why was it started? Why did so many states establish state schools for the deaf and blind in the 19th century? The list of such questions associated with education policy and finance is extensive.

Policy and practice in school finance have also been affected by history. Both have developed over time and are the products of the historical context from which they emanated. Most policy shifts in education have financial resource implications, and the major changes and necessary improvements are often not undertaken because of cost. The system of schools that exists today developed over time from the social and political eras of the past. The failure to understand the lessons of history leads to failure, while successfully moving ahead to the future requires knowledge of the past.

Litigation

Among the most influential shapers of education finance have been the courts. At strategic points along the development of the education system, the courts adjudicated watershed cases that altered state and local funding schemes. This influence has been seen mostly in state-level cases. While not precedent setting outside the state, select cases nevertheless had an impact in other states as political leaders moved to change laws before similar legal challenges arose in their state. The courts

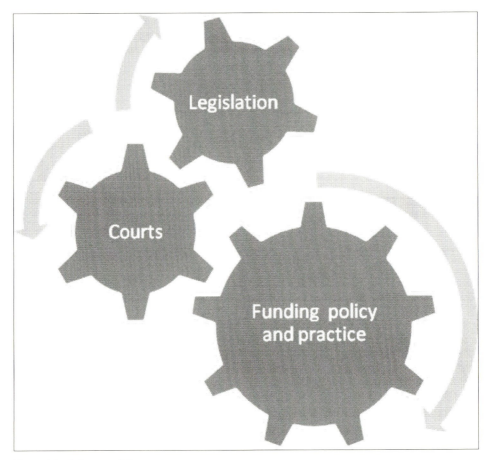

Figure 1.2   Influence of the courts.

have been at the forefront of groundbreaking legal interpretations of school funding approaches that are disallowed and that are viable (Alexander and Alexander, 2009).

As stated in an earlier section in this chapter, the U.S. Constitution is a central influence on the public schools. The courts routinely judge the balance between the propriety of the government in its funding of schools and the impact of those government decisions on the rights of individuals and groups. In those cases where the courts find the government acting in ways that violate individual rights or the equal treatment of various groups, the court steps in to correct the injustice.

Litigation in school finance has come a long way in carving the policy landscape for school funding. Among the policy questions addressed by the courts have been: issues about the appropriateness of taxing individuals in order to pay for educating the community's children; defining the appropriateness of spending tax revenue on various education and education–related functions; who should benefit from public education; what is a fair system of funding; how much funding is sufficient; and what financial considerations are reasonable for special populations?

In its role as interpreter of the law the courts have often demolished political logjams around school finance. Because of the scale and expense of many education funding disputes, the legislature is incapable of coming to consensus, even in the face of obvious needs for correction. When this happens aggrieved parties seek redress through the courts, which time and again have moved the policy agenda forward or overcome the entrenched politics of the times.

Economics

The United States spends about 4 percent of its gross domestic product (GDP)— the sum of all goods and services produced by a nation during one year—on PK–12 education (Congressional Budget Office [CBO], 2009). Elementary and secondary education accounts for an annual expenditure of over $650 billion as part of calculating that percentage. The revenue considered in developing these figures comes from local, state and federal sources. These revenues come from taxes, which are dependent on the assessed value of property, personal income and profits from business, commerce and the gain on investments of individuals.

Businesses and individuals are subject to the business cycle, i.e., the advancing and declining fluctuations of profit and loss associated with increasing and decreasing activity in business, industry and commerce. The schools are also affected by the various phases of the economy. As profits increase, revenue to taxing authorities rises; for example, more retail business equates to higher sales tax revenue; more construction adds to the base of taxable property; and rising income means more revenue collected from this source (Krugmen and Obsfeld, 2008). Often, during periods when the business cycle is rising, increases to school budgets are made. This contrasts with periods of decline, lower profits and reduced tax revenue. In such times school budgets remain stagnant or are reduced.

The global recession that started in 2008–09 is the latest example of an economic downturn that resulted in major cutbacks to school funding across the nation. This bottoming of the business cycle translated into smaller amounts of spending for school operations and capital projects. Most school districts around the country had to lower spending during this time, and in an undertaking like education, which is so labor intensive, it meant reducing staff or layoffs.

This recession, like all recessions, had its own special characteristics. One such aspect was the decline in property values, which in some communities was precipitous. In addition to being a major blow to the economy overall, it held special importance to school districts that rely on property taxes for a significant part of their funding.

Schools are an essential part of the economy in that schools develop what economists call "human capital," or the skilled people and intellectual power that support the economy. In turn, schools need a sustainable economy that can produce enough wealth to share in the support of schools. The interrelationship of economics and education is closely linked in modern developed nations. The economy is central to securing money for schools, and schools are integral to a viable economy.

Picture 1.2   Economic markets affect school funding.

Practice

The methods of funding schools and the things that are funded, like many of the other topics covered in this chapter, developed over time. Today we have a system of school finance in each state that has grown ever more sophisticated in terms of determining need, garnering resources, distributing funding, measuring equity in like and different student circumstances, using funds prudently, accounting for the use of resources and evaluating the effectiveness of the funding in terms of the mission of the schools. This is a far cry from the early days of schooling in America, when a group of families would get together, scrape off a corner of a fallow field, build a rough cabin as their schoolhouse and chip in to hire an itinerant teacher for a couple of months each year.

Today the amount of funds, the variety of sources, the array of education programs and the systems for keeping track of it all are vast and sophisticated. What might seem like a chaotic amalgam of 18,000 local education agencies and 99,000 schools has been organized into a semiautonomous whole that we call American public education. While hardly perfect, a system for answering important policy questions and informing education policy development is in place today. The data available from the PK–12 system serve many users from the local school board to the U.S. Congress. Much of the credit goes to dedicated school and government workers at the local, state and federal levels who have built a methodology for reporting and accounting for the billions of dollars spent each year.

Through these efforts financial reporting across the nation contributes to coherent policy development and analysis. Budgetary approaches at each level of government are more effective and accounting procedures more uniform. The direction and control of resources flows through the system of school finances in ways that are transparent and supportive of our democratic society. Remarkably, relatively little corruption exists across the country with school funding, in part because of the accounting methods built over the years.

The governmental and school portions of the generally accepted accounting principles (GAAP) have functioned well and are regularly refined to render even better accountability and data for policy makers and school leaders. Thus, through a developmental history, the cooperation of various levels of government and the commitment of policy makers, education leaders, finance scholars and accounting professionals, our modern-day system of school finance has come together.

Summary

In this chapter some, but not all, of the influences on school finance policy and practice are introduced. These topics are highlighted in an attempt to help the reader perceive the nexus among the range of factors that influence and shape the school finance system extant today. The chapter is offered to build a context and serve as a pre-organizer for the more in-depth coverage throughout the text. Subsequent chapters provide more detail on the issues raised here and present multiple dimensions to the other topics of importance.

It should be clear by now that the study of education finance policy, politics, theory and practice is more than an investigation of ledgers and obscure accounting methods. Examination of the topics in this book is intended to reveal the array of aspects that make up this dynamic area of pre-collegiate education. Education finance is multidimensional, and the purpose of the text is to broaden the understanding of current and future policy and school leaders.

Chapter 2 provides a more in-depth look at how the school systems in the United States developed over time. Special emphasis is given to the challenges associated with funding the schools. The interplay of politics, policy and practice is seen through this historical perspective.

References

Alexander, K., and Alexander, M. D. (2009). *American public school law* (7th ed.). Belmont, CA: Wadsworth, Cengage Learning.

Checchi, D. (2008). *The economics of education: Human capital, family background and inequality.* London: Cambridge University Press.

Congressional Budget Office (2009). Issues and options in infrastructure investment. Retrieved from http://www.cbo.gov/ftpdocs/91xx/doc9135/AppendixA.4.1.shtml.

Krugmen, P. R., and Obsfeld, M. (2008). *International economics: Theory and policy* (8th ed.). London: Pearson Education.

McDonnell, L. M. (2009). Repositioning politics in education's circle of knowledge. *Education Researcher, 38*(6), 417–427.

National Center for Education Statistics (2010). *Digest of education statistics, 2010.* Washington, DC: National Center for Education Statistics.

Neustadt, R. E., and May, E. R. (1986). *Thinking in time: The lessons of history for decision-makers.* New York: Free Press.

Oxford English Dictionary (2009). Retrieved from http://dictionary.oed.com. (Para 4).

Stone, D. A. (1997). *Policy paradox: The art of political decision making.* New York: W.W. Norton.

Wirt, F., and Kirst, M. (2001). *The political dynamics of American education* (2nd ed.). Richmond, CA: McCutchan Publishing Corporation.

Historical Perspectives on School Finance   2

Aim of the Chapter

THIS CHAPTER STRIVES TO HELP THE READER UNDERSTAND how the mechanisms in place today, used to finance the pre-collegiate public education system in America, developed over a long period of time and have roots that are surprisingly deep. The chapter provides a brief historical overview, explores those roots and draws connections between the actions of past generations to establish and maintain schools and the policies and practices of funding schools today. In some cases watershed events are used to capture the essence of the historical time, and in other cases historical anecdotes are presented in an attempt to fill in the context and texture of historical events.

Introduction

The course of history is rarely a straight line from event to action or idea to implementation. Sometimes it is difficult to make the association between how and why institutions that function today came into being. Additionally, one is reminded that revisionism is a frequent companion of historical interpretation, and we are no less susceptible to myth and legend today than past generations. All are encouraged to investigate cited references and original source documents for themselves in order to form a personal opinion about how things have come to be today.

It is important to keep in mind the overshadowing significance of cultural context to the understanding of history. Human events occur within a frame of time and place, but they also happen within and between cultures. History has deeper meaning to the extent one has insight into the culture of the society under consideration.

This chapter makes a broad sweep of American history and the development of publicly financed schools. Four major eras are examined: the colonial period from the sixteenth to the eighteenth century; the age of nation building after the American Revolution to the Civil War; the time of Reconstruction through World War II; and the last half of the twentieth century from the Civil Rights Movement and Cold War to the present. These historical ages each involved seminal events that greatly shaped the direction and growth of the United States and its society. It is to be expected, then, that one of the most significant institutions in America, its public schools, would also be greatly influenced by this history.

Historical Context

Since ancient times in societies around the world, humans have used the school as an instrument to transmit culture from one generation to the next. Schools, defined as groups of individuals engaged in a set of common learning or curriculum, emerge among groups of humans when it is no longer practical for the family unit to transmit the essential elements of the culture to the succeeding generation. Sometimes the school emerges because families are too busy trying to survive to take the time to teach youngsters. In other circumstances the knowledge to be transferred is too complex and requires a long period of study and a specialist to teach it (Ramirez, 2009).

By and large, however, schools were not the main source of cultural transmission until much later in human history, about the nineteenth century. While we can see examples of schools going back to ancient China, Egypt, Mesoamerica and Greece, for most children education consisted of learning their role in society from their father or mother, as gender dictated. Beyond the home, religious institutions, through ceremony and direct instruction, passed on important cultural knowledge.

The apprenticeship served as a common source of special training related to economic survival. Many examples of the apprenticeship included a form of indentured servitude, where children were apprenticed to a neighbor or family in a local village for a period of time in exchange for the child's labor and the promise that the child would learn a useful skill or trade and be readied for adulthood in the larger society.

Schools develop when a society becomes more complex and has a need to create large numbers of specialists or transmit technical information that the home neither has nor is able to convey. For example, clerics, large-scale builders, and military and naval officers from times past are frequently educated in what we today would call a school. As a trading or mercantile economy emerges within a society, the formation of guilds and their specialized schools related to manufacturing or trade will form.

In the United States, schools developed for similar reasons, and as the needs of society changed, so did the need for schooling. It is important to keep in mind as we view these historical periods that our point of view is focused on historical events as they influenced the development of publicly supported schools; thus, this limited narrative will spotlight how schools were financed and related historical events.

The Native Peoples, European and African Culture in America

Many things motivated those who ventured across a vast ocean in small wooden ships to what would later be called America. Some sought quick fortune and glory and came with no intention of staying. Some came to build empires. Others came to escape oppression and pursue spiritual freedom. Countless numbers came for the chance at a better life through access to free or cheap land. Many others were brought to the "New World" in chains to be the hard labor that would in fact build a new world from the vast bountiful land. And, when early explorers and settlers from Europe and Africa arrived, they encountered an exotic native people who already had an established way of life.

Native peoples valued their styles of education, and their children learned about their cultural heritage, spiritual beliefs of their people and the means of survival in their physical environment. This learning took many forms and served them well, as evidenced by the proliferation of native groups throughout the Americas and a people who had occupied much of the land for eons. For the most part education was gender specific, traditional and apprentice-like. Sons followed fathers and daughters followed mothers, and the more complex societies had more levels of specialization.

Each of these groups held at least one thing in common; they each originated from a unique culture. And each of these groups, like all groups, strove to preserve and perpetuate their culture. For the Europeans it is essential to understand the historical setting from which they emerged during the colonial period. Two significant events helped to shape the motivations and actions of Europeans in the Americas and significantly influenced the formation of schools during the early colonial period: the first was the impact of Johann Gutenberg's invention of printing technology in the middle of the fifteenth century, because his technology eventually made it possible for the average person to own or have access to books; the second was the religious strife in Europe, notably, the Protestant Reformation, and the Catholic Counter-Reformation.

The movable type printing press technology made it economically possible within a generation of its development to afford the average person access to books. Prior to this breakthrough, books were rare objects collected mostly by the nobility, the church, the synagogue, a wealthy merchant, and the handful of universities in existence at that time; most people might glimpse a book while at church but have no need to own one. With access to books, reading becomes a more practical skill; the subject matter of books is transportable and expands to more temporal and practical knowledge.

The Protestant Reformation, among other things, set in motion a fierce competition between Roman Catholic-aligned nations and Protestant-aligned nations.

This rivalry migrated from Europe to the Americas along with the Europeans. This was a competition not only for wealth, power and turf, but also for souls. For the Europeans from the Iberian Peninsula, Spain and Portugal, the Catholic versus Protestant hostilities marked yet another chapter in a very long (seven-hundred-year) saga of religious conflict as they shifted at the end of the fifteenth century from concerns about Muslims to concerns about Protestants.

Thus, from the very beginning of the European and African migrations, a struggle for cultural dominance compelled the transmission of religious knowledge from one generation to the next and one group to the other. In this regard the European migrants proceeded with two objectives in mind: first, that they establish in the New World the institutions that would recreate the culture they had left behind; and second, that they assert their culture, i.e., language and religion, on the African and native populations they encountered.

It is apparent from the outset that part of the goal for the European explorers and settlers, as articulated in the enabling charters that authorized and funded their ventures, was that they had a requirement to impose their culture, particularly religion, on the native population. Consider the first entry in *The Log of Christopher Columbus* in 1492:

> In the Name of Our Lord Jesus Christ Most Christian, exalted, excellent, and powerful princes, King and Queen of the Spains and of the islands of the sea, Our Sovereigns: . . . Your Highnesses decided to send me, Christopher Columbus, to the region of India, to see the Princes there and the peoples and the lands, and to learn of their disposition, and of everything, and the measures which could be taken for their conversion to our Holy Faith. (Columbus, 1493/1987, p. 51)

Similarly, the First Charter of Virginia, granted in 1606 by King James I of England, more than one hundred years later ordered that the colony work on:

> propagating of Christian Religion to such People, as yet live in darkness and miserable Ignorance of the true Knowledge and Worship of God . . . in time bring Infidels and Savages living in those Parts, to human Civility, and to a settled and quiet Government. (Szasz, 1988, p. 46)

The Virginia charter also included a provision for financing schools to achieve these goals and to ensure the children of English colonists were protected from the "barbarous influences of the land." The method, so common throughout the historical record, was to set aside land, to be exploited or sold, the proceeds of which were to support the educational enterprise.

The system of Kings or Queens granting charters, or exclusive franchises, to individuals or corporations who in turn derived wealth from the land and shared the profits with the monarch, was long established. What can be seen from the charters and other enfranchising edicts is that these enterprises were also interested in establishing the culture of the sponsoring nation in the new land. And, schools were seen as a vehicle for accomplishing this goal.

Picture 2.1   The Mission used education to transmit culture to European colonies.

For the African immigrants, ripped from their homeland by kidnappers and slavers, the devastation to their culture was almost complete. Thrown together with individuals from various areas of Africa and different language groups, those that survived the Atlantic crossing were further dispersed to far-flung territories throughout the Americas. When they arrived at their final destination they faced an existence that ranged from brutal survival to the lowest caste in society. Their status varied widely depending on where they ended up and whose colonial territory they were in. However, within a few generations, a unique African American culture emerged in every country and region to which they were brought.

Education for the African population also varied widely. Most of the people ended up on plantations. The plantation was organized to produce large quantities of cash crops, and like the Spanish mission, designed to be self-sustaining. As a result, a division of labor and specialization within the labor pool was required. While most of the labor was assigned to the planting, growing and harvesting of crops, builders, butchers and blacksmiths, along with hostlers and leather crafters, were part of the variety of skilled and semi-skilled workers to be found on plantations. Often, plantation owners would contract with individuals from outside the plantation for the labor these skilled workers provided. The transmission of vocational skill knowledge required a system of organized learning, the most common being the apprenticeship.

A big part of the subjugation of the African population involved acculturating them to the European world view, a major part of which was conversion to the

Christian faith. This endeavor posed a moral struggle within the white population of the European colonies in North America, and later the emerging nation, that went on for centuries. On the one hand, religious leaders like Bishop Edmund Gibson of the See of London in the early 1700s urged slaveholders to instruct their charges in the Christian faith and encouraged missionaries to tend to this flock. On the other hand, the colonies—and later states—enacted laws prohibiting the assembly of slaves; the establishment of schools for any person, free or slave, of African descent; and the teaching of reading and writing to them (Katz, 1969).

Not withstanding such laws and customs, there are numerous cases of schools being established for African Americans, free and slave, throughout the colonial and postcolonial era. The Society for the Propagation of the Gospel in Foreign Parts was a major force in this regard. For example, in New York in 1704, Elias Neau started a school for black slaves. It was later disbanded after an uprising among the black population in the city. Dr. Bearcroft reports in 1744 of his school in Charles Town, South Carolina, that educated a select number of black slaves in the scriptures and to read and write. His goal was to send the graduates back to their areas to serve as missionaries to their fellow slaves.

The Quakers, members of the Society of Friends, encouraged the conversion and education of the non-white population. In 1770 in Philadelphia, a school was established for black and mulatto children, money was allocated and a teacher was hired. By the early 1800s, colleges began to emerge with the objective of training people of color who in turn would serve as schoolmasters and clergy throughout the region (Katz, 1969).

Colonial Roots

Despite the zeal of a few individuals and the edicts of European monarchs, a system of publicly financed schools would not emerge for many generations. But, some of the roots of what was to evolve into the American system of mass education did start to take hold. Cubberley (1948) points to the case of New England, and specifically the Massachusetts Bay Colony, as the most influential wellspring of our modern-day system of American schooling. He notes that the early settlers, the Puritans, were a well-educated group of mostly middle-class means. Their interest in education for religious and practical reasons was well established. As a case in point, within less than two decades of landing at Plymouth Rock, they established a college, Harvard, in 1636, to ensure future generations of ministers.

Initial efforts by the colony's General Court, its legislative body, to encourage parents to assume the rudimentary education of their children were determined to be inadequate. Thus, in 1642 the colony established the "Ye Old Deluder" law, an effort justified to counter the work of Satan, who was perceived to want people ignorant of scripture. This law required that all children learn to read, the first law of its kind in the English-speaking world according to Cubberley (1948). It also

---

Text Box 2.1 The influence of the church mission.

The church-based mission was yet another means through which migrating Euro-peans attempted to transplant their way of life to new lands. This institution often incorporated a religious, economic and military purpose into a comprehensive system of self-contained settlements that helped the Europeans gain a foothold in frontier territories throughout the Americas. The mission sought to become a self-sustaining operation that duplicated European beliefs and technology. Thus, agriculture, animal husbandry, manufacturing and mining were among the crafts and skilled trades taught to the native people and the children of colonists in the schools that functioned in the mission. Of course, the teaching of a Christian-centered life was also part of the mission schools, and in some cases selected native and white individuals were groomed for higher forms of education and roles as future missionaries in new territories.

The Society of Jesus (Jesuits), the Order of Friars Minor (Franciscans) and the Order of Friars Preachers (Dominicans) were the most prolific builders of missions in the Americas during the 16th, 17th and 18th centuries. Many Protestant groups built missions as well, and the competition for souls remained fierce. From New France, throughout the Great Lakes and down the Mississippi river; to the English colonies such as Maryland; and to New Spain, from Georgia to Florida to Texas to California; the missions sprang up as the vanguard of European colonization. Education was at the heart of the purpose of these institutions, and the financing for sustaining the mission came from charitable donations, the labor of native people and the exploitation of the land.

Many of these missions failed. Often they were not able to achieve the sustain-ability they desired or they lost their strategic significance. In other cases, without a military garrison for protection, they could not survive attacks by native groups, whether a revolt by the local population or from a competing neighboring tribe. Some missions thrived and became the cornerstone of our modern cities: from Green Bay, to Montreal, to Natchez, to New Orleans, to San Diego and San Francisco (Moore, 1982). The mission school for the native people and conflict between Protestant and Catholic dogma would extend well into the twentieth century (Prucha, 1979).

---

required parents to account to town leaders regarding the status of education of their children, subject to fines for failing to render the accounting.

After five years the General Court determined this law, too, was inadequate and substantially revised it. The 1647 law provided that "every town of fifty families appoint a teacher of reading and writing and provide for his wages in such manner as the town might determine; and, every town having one hundred households must provide a grammar school to fit youths for the university, under penalty of five pounds (later increased to twenty) for failure to do so." The law of 1647 is significant for the rationales used and the precedents it set:

1. The universal education of youth is essential to the well-being of the State.
2. The obligation to furnish this education rests primarily upon the parent.
3. The State has a right to enforce this obligation.
4. The State may fix a standard which shall determine the kind of education and minimum amount.
5. Public money, raised by general tax, may be used to provide such education as the State requires. The tax may be general, though the school attendance is not.
6. Education higher than the rudiments may be supplied by the State. Opportunity must be provided, at public expense, for youths who wish to be fitted for the university. (Cubberley, 1948, p. 366)

A modern day school administrator looking over the above list might be tempted to add that it represents the first example of an unfunded state mandate to a local school government.

This model for establishing schools was quickly adopted in the neighboring New England colonies, which had unique social capital based on their origins, makeup and organization. The model was also influential years later as the United States expanded beyond the territory of the thirteen original colonies and started to add new states. For example, John D. Pierce was a transplanted New Englander who became Michigan's first state superintendent of public instruction in the early nineteenth century (Cremin, 1980). He was the architect of a comprehensive state system of education, designed in part on the traditions of the New England schools. But it is a mistake to imagine a direct line of development from colonial New England to our modern-day education system.

The Middle Atlantic States were dominated by parochial schools and charity schools for the poor. The spiritual admonition to parents to teach their children to read so they could follow the scriptures was not as strong as in New England. The South relied on pauper schools for some indigent children, with private schools, church schools, tutors and study abroad as options available for the elite classes. Poor whites in the South could not afford the private schools and only rarely had the opportunity to attend charity institutions (Anderson, 1988). Colonies that had multiple religious denominations and sects within their borders—for example, New York, Pennsylvania, and New Jersey—found it impossible to gain a consensus regarding a mandate for a uniform approach to education as seen in much of New England. This diversity, while at once a strength, also contributed to the challenges in later years of building a consensus around a publicly funded system of schools.

Building a New Nation

After the defeat of the British in the American Revolutionary War, the leaders of the land turned their attention to the unprecedented task of building a republic comprised of the vast territory of America. The principles upon which the new republic would be built embodied several key suppositions:

1. The right to propose measures and policies.
2. The right to discuss proposed policies and measures.
3. The right to decide issues at the polls.
4. The obligation to accept decisions duly made with no resort to force.
5. The right to appraise, criticize, and amend decisions so made. (Peterson, 1975, p. 10)

In order for this democratic system to function, a critical lynchpin was required—an educated citizenry. Without education, it was argued, ignorant mobs would rule or be led by conniving schemers, and the liberty won on the bloody battlefields would be lost.

The 13 colonies, now states within a confederation, had very different ideas about how to approach the educational needs of their respective citizens. But key leaders, founding fathers among them, understood that the need existed. John Adams argued for "laws for the liberal education of youth." Thomas Jefferson, in a letter to his friend Colonel Chauncy, stated it clearly, "If a people expect to be ignorant and free, they expect what never was and never will be." Jefferson also proffered strong arguments regarding the funding of public education, "the tax which will be paid for this purpose is not more than the thousandth part of what will be paid to kings, priests, and nobles who will rise up among us if we leave the people in ignorance" (Alexander and Alexander, 2004). Benjamin Franklin, Benjamin Rush and Daniel Webster were also enthusiastic advocates of education for the masses as a route to creating a unique American people (Cremin, 1970).

The funding of schools among the states in the nation consisted of a laundry list of financing schemes devoid of a systematic process. Funding sources included:

○ rate bills, where parents were taxed based on the number of children attending school;
○ the proceeds of government land sales or the rent of government land;
○ timber and mining leases on government land;
○ bequests of estates to schools;
○ proceeds from public utilities, e.g., toll bridges, ferries, toll roads, self-sustaining farms, mills;
○ donations;
○ endowments;
○ lotteries
○ subscriptions;
○ charters granted by local governments to religious organizations;
○ and, in a perverse twist of irony, there are even examples of the proceeds of the sale of slaves being used to fund common schools. (see the example of the Florida territory in Katz, 1969, p. 337)

Despite the admonitions of the visionary leaders of the day the general availability of publicly funded education was not to be seen until the middle of the nineteenth century, and much later in many parts of the South. The commonly held perspective at the end of the eighteenth century was that basic education was essentially a family function or something the church did. Additionally, the nation and individual states were virtually bankrupt after the Revolutionary War. Notwithstanding these economic conditions and prevailing attitudes, some states did move forward to "encourage" schools.

Governor George Clinton of New York State in 1795 succeeded in passing a law that appropriated two thousand pounds for the support of schools. The money was to go to towns willing to tax themselves for the maintenance of schools. New York City was allowed to use some of the funds for its charity schools. This soon led to some of the Catholic schools in the city asking for their share of the funding (Cremin, 1980).

By 1812, New York had enacted a school law that started to look like something we would recognize today. The law provided for a state superintendent with duties to develop plans to better manage schools and their resources. As Cremin (1980) points out, "the new law was patently intended to erect a system." The three tiers of the system consisted of: local school districts created by towns, and responsible for the maintenance of buildings; the towns, responsible for hiring and supervising teachers; and the state, responsible for assisting the districts and distributing money from the permanent school fund.

The Continental Congress passed the Ordinances of 1784, 1785 and 1787 to serve many purposes for the newly forming nation. The ordinances provided the framework through which the territories of the "Northwest" could prepare themselves to become states and enter the union. The Northwest Ordinance of 1787 referred to territory northwest of the Ohio River, for example, present-day Ohio, Michigan and Illinois. When a territory achieved the requisite population and had established functioning institutions like a legislature and courts, it could petition the national government for admission to the union as a full-fledged member. Among the institutions promoted by the Ordinance of 1787 were schools: "Religion, morality, and knowledge, being necessary to good government and the happiness of mankind, schools and the means of education shall forever be encouraged" (Our documents, 2012, sec. 14, art. 3).

Note that these laws predate the ratification of the U.S. Constitution in 1788, which is silent on the matter of education even to modern times. Furthermore, the ordinances did provide a means of financing, at least in part, necessary institutions. The method used to raise money was the proceeds from the sale of lands.

The Ordinance of 1787 provided that the new territory should be divided into townships of six miles by six miles square. The townships were to be further subdivided into sections of one mile square, so each township had thirty-six sections. The one square mile section was equal to 640 acres. Section 16 of each township

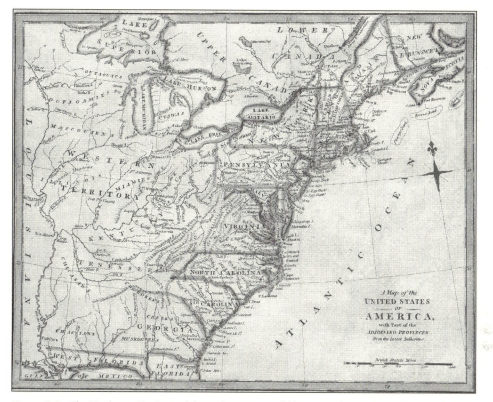

Picture 2.2   The Northwest Territory of the early 1800s would become Ohio, Indiana, Illinois, Michigan and Wisconsin.

was reserved for schools. The proceeds of the sale or lease of section 16 lands were put in the "Permanent School Fund" to help the territory or new state begin to finance its education system.

The ordinances are very significant in that they influenced the development of all the states admitted to the union thereafter. They were designed to entice families, i.e., permanent settlers as opposed to single men. And, it can be argued they also influenced how schooling and the government's role in it were viewed in many of the original states as the old states reflected on the progress of public education in the new states. The ordinances established several important precedents:

1. An educated citizenry is essential to a republican form of government.
2. There is a legitimate role for government in education.
3. Education is fundamentally the domain of the state government.
4. The national government should promote and support education.
5. School finance was essentially a state matter.
6. Schools should be part of a state system.
7. The concept of publicly supported nonsectarian schools was viable.

The process of providing land grants to newly forming states to help them start public institutions of all types would continue beyond the nineteenth century and include New Mexico and Arizona as late as 1912. In total the states received 77,630,000 acres of land for common schools purposes. Over time Congress became more prescriptive with its land grant legislation in an effort to avoid some of the corruption and mismanagement of the federal land grants (Tyack, James and Benavot, 1987). It also became more exacting with regard to the education requirements for territories that wished to enter the union. So it seems natural in retrospect that the U.S. Congress in 1867 established the Bureau of Education in the Department of the Interior.

The growth of the nation during this era was astounding. The U.S. population, according to the first census in 1790, was four million. By 1870 the nation's population had exploded by 1,000 percent to forty million, fueled by the torrent of immigrants from Europe. One result of the massive influx of foreign-born new Americans was an intensified emphasis on schools as a public institution necessary to perpetuate American ideals and the American way of life.

Many of the immigrant groups brought their cultural institutions with them. Thus, churches and parish schools grew and spread along with the immigrant populations. Many native-born citizens reacted by turning to public education as a necessary vehicle for "Americanizing" the immigrants and their children.

To a large extent the common school movement of the middle of the 1800s was given impetus by the growing foreign-born population, the industrialization of the economy and the tremendous growth of the cities that begin to emerge as great concentrations of political power. Schools could not be built fast enough to accommodate the burgeoning school-age population.

Tyack (1974) explains how the Lancasterian plan, or system, was a widespread model used to educate the masses of children in urban schools. This approach saw a teacher or two with as many as 200 children organized into subgroups of peer-age learners who were tutored by older students, called monitors. From a school finance perspective this was a highly efficient model. However, parents and education reformers of the day were critical of this approach, which they saw as inadequate.

John Philbrick in 1848 is credited with introducing the Prussian model, an innovation that was to catch on like wildfire across the nation. In Massachusetts, Philbrick opened the new Quincy School, a four-story structure for seven hundred students who were organized into twelve classrooms, each with its own teacher and about fifty-six students. The children were divided by their tested proficiency and organized into "grades." An administrator, usually a male, supervised the twelve female teachers. Compared to other models of mass education of the time, this method was much more costly because of the increased labor costs and specialized facilities.

The emphasis during this era of American education was on building capacity and in systematizing the array of publicly funded schools and programs popping up like mushrooms across the cities and countryside. School finance schemes, along with standardized curriculum and testing, uniform facilities, bureaucratic rules and teacher training were methods of achieving a coherent system of public schools.

The Post Civil War to World War II

The concept of common schools that were publicly financed had taken firm hold in the Northeast, Middle West and the new states of the West by the time of the Civil War (1861–1865). Most of these states were engaged in building or expanding their education systems. The Free School Societies, labor unions and philanthropic organizations advocated mightily for publicly funded and run schools. The work of James Carter, Horace Mann, Henry Barnard and other education reformers of the early nineteenth century was paying off.

Normal schools or teacher training institutions were emerging around the country, and efforts to professionalize teaching and the administration of schools were starting to move beyond the discussion stage, although, these institutions and professional standards would not take hold until much later in the century. Teenage girls and young women became the labor force for this growing system as teaching became a "respectable" profession for a single, young woman. They were also paid much less than the male teachers who had dominated teaching in the earlier era.

The southern states chose a different path. The common school movement was not popular in the South until the Reconstruction Era, and even then it was African Americans who advocated most vigorously for free schools. Anderson (1988) quotes W. E. B. Du Bois as saying, "Public education for all at public expense was in the South a Negro idea" (p. 6). The newly enfranchised citizens understood that participating in the democracy meant voting and voting required literacy.

The conditions for readmission to the Union for the former rebel states required that the state draft a new state constitution, which was to be scrutinized by Congress. As with the new territories of the West petitioning for statehood, the former Confederate states had to organize their governments within prescribed parameters in order to succeed in their petition for readmission to the Union. Among the required institutions for the southern states was a public education system. So by 1868, Alabama organized a public school system; Mississippi and Georgia followed suit in 1870.

But support for publicly financed nonsectarian schools was not universal. For example, the *Houston Weekly Telegraph* in 1868 was a critic of taxes for schools. Collecting established taxes proved difficult, and many systems were underfunded to the point of not being able to function. The Alabama system is an example where planters discouraged schools as a tax on them and a negative factor on their labor force.

The Freedmen's Bureau, established in 1865, was a federal agency with a mission to assist the newly freed slaves. It did much to advance public education for

African Americans during Reconstruction, as did white missionaries from the North. Reconstruction and the imposed governments backed by federal troops were not popular among whites. So while the form of a public school system was adopted in many of the former Confederate states, the substance of such systems was still many years away.

Where the public education systems did survive, contrasts in funding and services for whites and blacks were stark. Bullock (1970) documents expenditures across a range of categories and shows growing disparities between the races as the federal influence diminished. He considers the percent of black and white children in school; the per capita dollars spent on each group; and other items like teacher salaries for black and white teachers.

Many things impeded the establishment of public education systems in the South, but two are of particular note. The presidential election of 1876, between Republican Rutherford B. Hayes and Democrat Samuel J. Tilden, ended in a tied vote with disputed returns from Florida, Louisiana and South Carolina. A compromise was struck to give the election to Hayes under the condition that federal troops be withdrawn from the South and Reconstruction ended. As African Americans lost political power under the new arrangement, their access to schools diminished.

A second major event to adversely impact the education of African Americans was the 1896 landmark U.S. Supreme Court decision of *Plessy v. Ferguson*. This decision enshrined the doctrine of "separate but equal" in the area of public facilities for blacks and whites. Combined with the loss of the federal influence, this decision crystallized the development of a dual system of education in the South.

Blacks continued to fight for education and in many cases used their own money to fund schools or supplement the meager amounts received from the responsible local government. Anderson (1988) points to the Georgia Education Association, 1865, as an example of a black–organized, –run, and –financed effort at a school system. He also describes efforts by African American agricultural workers to negotiate education clauses into labor contracts. In some cases blacks were saddled with school taxes but provided no schools and had to pay for their own schools, in essence sustaining a double tax.

Not all whites were opposed to education for blacks. Southern industrialists saw education as a way to prepare a disciplined, racially stratified labor force for the factories; missionaries continued to advocate for black education and fund and staff schools; and northern philanthropists donated much for facilities and teachers. But the pattern of gerrymandered school districts and taxing districts combined with the dual system of public education served to minimize or outright deny public education to African American children. Not until the latter part of the nineteenth century did poor whites in large numbers also begin to support the idea of universal free education. Thus the private academy, church school and charity school were left for the small number of poor white families who had access to or could afford education.

Struggles for Access

The effect of racist attitudes backed by government policies was not limited to the South. A central judicial precedent that girded the *Plessy* decision was found in *Roberts v. City of Boston* in 1850, which upheld the legality of segregated schools in that city (Tyack et. al., 1987). The authors also explain that:

> Blacks in the North and South, and Chinese in California, and other suppressed groups like Indians and Hispanics, often learned that the public school was common for Whites, not for people of Color. (p. 17)

Katz (1969), editor of *History of schools for the colored population*, captures numerous excerpts from government reports and records of the time that give a clear glimpse of the era. Some examples follow. Illinois determined that since its constitution of 1847 restricted the right to vote to white males, by extension the school laws only applied to whites. In 1868, Newton Batemen, state superintendent of Illinois, exhorted the general assembly to remove the word "white" from school law and open the schools to all children. The superintendent of schools for Chicago reported on the city's failed attempt at a school for black students and the need to integrate these children with their peers. The school was subsequently abolished by the legislature in 1865. He extolled the success of the integration of schools thereafter.

The New York legislature passed a school law in 1841, revised in 1864, that authorized any school district to establish separate schools for people of African descent. The intent regarding the financing of these schools was made clear:

> and such schools shall be supported in the same manner and to the same extent as the schools supported therein for White children; and they shall be subject to the same rules and regulations and be furnished with facilities for instruction equal to those furnished to the White schools therein. (Katz, 1969, p. 361)

Out West

The superintendent of public instruction in California reported in 1867 that there were 709 Negro children between the ages of five and sixteen, and that four hundred of them were being educated in 16 segregated schools. He is quoted as saying "the people of the state are decidedly in favor of separate schools for colored children" (Katz, 1969 p. 328). The revised school law of California in 1866 was clear on the subject:

> Sec. 57. Children of African or Mongolian descent, and Indian children not living under the care of White persons, shall not be admitted into public schools, except as provided in this act: Provided, That, upon written application of the parents or guardians of at least 10 such children to any board of trustees, or board of education, a separate school shall be established for the education of such children, and the education of a less number may be provided for by the trustees in any other manner.

Sec. 58. When there shall be in any district any number of children, other than White children, whose education can be provided for in no other way, the trustees, by a majority vote, may permit such children to attend school for White children: Provided, That a majority of the parents of the children attending such school make no objection, in writing, to be filed with the board of trustees.

Sec. 59. The same laws, rules and regulations which apply to schools for White children shall apply to schools for colored children. (Katz, 1969, p. 328)

The long struggle over assimilation of Native Americans into the Eurocentric culture saw an increase in the use of government-funded and -run boarding schools in the nineteenth century (Adams, 1995). In the Southwest and California, the Spanish Mission had a centuries-old history of that effort.

Population Explosion

Nationally, enrollments grew 100 percent between 1870 and 1898, and 71 percent of five- to eighteen-year-olds were in school (Tyack, 1974). By 1882, for example, there were 63,500 students in the Philadelphia school system. The superintendent of schools oversaw ninety-two schools. Issues of planning, pupil accounting and facilities were foremost in the minds of policy leaders. The urban population explosion forced school leaders to focus on the efficiency of their education systems.

The high school as an option for all youth also started to take hold in the late nineteenth century. In 1880, the number of public high schools surpassed that of private high schools or academies. However, funding schemes at the time, through "scholarships" and other public support to private schools, obscure the exact point at which the public high schools took the lead.

Tyack (1974) compiled some impressive figures to illustrate the growth of public high schools across the nation. In Chicago in 1894, there were only 732 seniors in high school out of a total school population of 18,500. Nationally only 10 percent of students graduated from high school.

Power (1970) marks 1821 as the origin of the American free public high school, which was first established in Boston, Massachusetts. Unlike the academy or the Latin grammar school that preceded it, the public high school focused on preparing boys and girls for the practical occupations of the day along with solid academic skills in English. Three key points of distinction for the public high school were that it was publicly funded, publicly controlled and open to all qualified students.

Table 2.1   The American high school developed slowly.

Year	Students	Percent of seventeen-year-olds
1870	16,000	2
1890	44,000	3.5
1900	95,000	6.4

Source: Tyack, D. (1974). *The one best system: A history of American urban education.* Cambridge, MA: Harvard University Press.

However, Peterson (1985) documents how "all" really means "some" in places like Atlanta, Georgia, where the struggle to establish high schools for black youngsters spanned many decades.

It took a court case in Michigan to overcome the legal impediments to establishing publicly funded high schools within the state public education systems around the country. The Kalamazoo Decision of 1874 determined that the Michigan High School Act of 1859 was indeed constitutional in Michigan. The court ruled that high schools were within the purview of public education and thus it was legal to spend tax dollars on such schools.

While not a precedent-setting decision, it proved very influential in mounting similar arguments in other states. Despite such efforts only 8 percent of the fourteen- to seventeen-year-old age group was in public high schools by the early 1900s, and many rural areas across the nation did not have high schools until after the First World War (Johns, Morphet and Alexander, 1983). Girls dominated the enrollment in the early days of the American high schools as school superintendents recruited the teenagers in order to collect the available local and state enrollment revenue.

The Cult of Efficiency

The U.S. involvement in World War I (1914–18) proved to be a revealing event for the nation in many ways. Among these revelations were the glaring discrepancies of the educational levels and basic literacy among conscripted soldiers. The adequacy and amount of schooling for the average soldier varied widely depending on the state he came from and whether he lived in the city, country or small town. Data collected during the war about education levels and attainment helped spark a new round of education reform in the first half of the twentieth century.

The end of the nineteenth century into the beginning of the twentieth was characterized by efforts to gain greater efficiency within state school systems of education, including attempts to depoliticize school districts and wrest control from ward boss politicians. The consolidation of rural school districts and small town, ward-based school districts advanced at a relentless pace despite resistance from local communities.

With the popularization of the automobile, transporting students to school by bus began. Standardization of curriculum, textbooks and program offerings, and centralization of decision making received much emphasis. As a result, larger school districts with larger schools became the trend. Control of financial resources moved toward the centralized school board and the school district superintendent.

Tax support for public schools throughout this era consisted mostly of locally raised tax dollars and some state monies. The flat grant, an equal amount of money allocated by the state on a per pupil basis, was a school funding innovation used by many states. The flat grant was an effort to improve on the lump sum approach previously used. Lump sum allocations tended to ignore the number of students to be served and thus created wild variations in the amount of state support available per child. The flat grant was a move toward equalizing funding for each child.

But these funding approaches proved inadequate and manifested wide disparities within and between states in such things as the length of the school year and day, class size, number and qualifications of teachers, teacher pay, school facilities and available textbooks and instructional materials. Thus, it followed that the educational results achieved by different groups of students also fluctuated greatly.

Fairness in School Funding

It was apparent to school reformers that upgrading and standardizing the education systems around the country would require new forms of school finance mechanisms. Johns et al. (1983) describe the development of what was to become the financing system found throughout the nation by the middle of the twentieth century. Ellwood P. Cubberley in 1906 published an influential study that raised the issue of the inadequacy of extant school funding schemes and led to the development of the more evolved foundation approach. Harlan Updegraff in 1921 proposed a sliding scale for state support tied to local effort and state equalization of funding. By 1923, George D. Strayer, Sr. and Robert M. Haig conceptualized what was to become known as the Strayer-Haig formula for the "Equalization of Educational Opportunity." They proposed a finance formula, which aimed to deliver sufficient financing to ensure each student a minimum satisfactory program through additional state support for low-wealth school districts that could not meet the minimum.

In 1924, Paul R. Mort helped to popularize the concepts of the Strayer-Haig formula. Mort's scholarship added to the definition of minimum program. Thus, many of the arguments, concepts and solutions to adequacy and fairness in funding schools, which to this day are studied in universities and debated in state legislatures and courts around the country, began in the early part of the twentieth century.

Federal Categorical Programs Begin

The early twentieth century also saw one of the first efforts at federal categorical grant in aid to states and school districts. The Smith-Hughes Act, passed in 1917, provided financial incentives to states to offer and expand vocational education programs at the high school level. Matching state funds were required as a condition of participation. Additionally, states had to establish governance systems specifically for vocational education, in many cases paralleling the already existing general education system.

Response to the Great Depression by the administration of Franklin Delano Roosevelt included many education-related programs run directly by the federal government. His New Deal initiatives like the Civilian Conservation Corps of

1933, Public Works Administration of 1933 and National Youth Administration of 1935 are examples. The National School Lunch Program was launched as a vehicle for price supports for farm products with depressed prices. But with the approach of World War II, industrial manpower needs coupled with massive conscriptions of young men for the war rendered many of these programs unnecessary.

The Civil Rights Movement and the Cold War

The aftermath of World War II saw many social changes in the United States. Among these were demands for change in the condition of schooling for minority groups. One of the most significant efforts in this regard was initiated by a group of Mexican American veterans in southern California who organized the Latin American Organization (Gonzalez, 1990). Their mission was the desegregation of schools in Orange County.

Resistance from the school board to their demands for an end to generations of segregated schools and a dual system of education precipitated a successful lawsuit brought in Ninth Circuit Federal Court in 1946, *Mendez et al. v. Westminster School District of Orange County et al.* The case rejected the doctrine of separate but equal solidified fifty years earlier in *Plessy v. Ferguson* (1896). *Mendez* and other similar cases helped to lay the groundwork for the landmark U.S. Supreme Court case *Brown v. Board of Education of Topeka, Kansas* in 1954, which among other things abolished the power of the state to legislate segregated schools.

The impact of *Brown* from a school finance perspective was enormous. Across the country disenfranchised minority groups brought successful lawsuits that resulted in a diminution of dual systems of education and a move toward equalization of resources within school districts. In some cases segregated schools were closed and new facilities opened; magnet schools with enriched program offerings designed to attract voluntary integration became a popular remedy. Federal judges took control of recalcitrant school districts and frequently ordered additional expenditures. Other expenditures increased, such as transportation, as children were bussed across town away from their neighborhood schools to achieve balanced integration in the schools.

It is difficult to overstate the significance of these school legal battles. The plaintiffs' success in linking the equal protection clause of the Fourteenth Amendment to the distribution of educational resources would serve challengers, including poor whites, of unfair school finance systems all over the nation. Beyond the raft of desegregation cases spurred by *Brown,* many successful school finance cases were brought regarding inequitable state funding systems that relied extensively on local property tax and thus delivered fewer resources to property-poor school districts. To this day litigation in state courts continues regarding the question of whether students are getting their fair share of funding. Many of these cases turn on the principles originally developed in *Mendez* and *Brown.*

The Proliferation of Federal Education Programs

No sooner had World War II ended than the Cold War started between the communist-aligned nations and the free market democracies of the West. One outgrowth of this conflict was a race for technical superiority in all fields of endeavor. The shock of the successful launch of the Soviet space satellite Sputnik on October 4, 1957, caused the United States to question, among other things, the adequacy of its education system. One outgrowth of this reflection was the enactment, in 1958, of the National Defense Education Act by Congress. This program infused relatively large sums of federal money, compared to earlier appropriations, into the states for improved education in a wide range of programs from science and mathematics, to foreign language, to teacher training, to college tuition assistance. Many states followed with their own reform initiatives to upgrade the public schools.

In the midst of the postwar prosperity of the late 1950s and 1960s, however, the glaring disparities that existed between the middle class and the poor caught the attention of the nation (Harrington, 1962). President Lyndon B. Johnson made his "War on Poverty" the centerpiece of his domestic policy agenda, and education was to be the vehicle to carry the poor to the American Dream. In 1965, Congress passed the Elementary and Secondary Education Act (ESEA) and began a federal education program that to this day provides funds for school districts with large concentrations of children from poor families.

As noted previously, federal actions in education often prompted state action. The attention to the plight of poor children at the national level caused state legislatures to evaluate their own state school finance systems. As a result, many states instituted supplemental funding programs to provide state-funded grants to school districts with high concentrations of disadvantaged children. In the name of compensatory education, states altered their school finance formulas to adjust funding allocations for school districts with large numbers of poor families.

The period from 1958 through the late 1970s saw a proliferation of categorical aid programs from both the federal and state level. School districts scrambled to write grant applications to receive funding for programs that ranged from early childhood education to community education programs for adults. But no program that emerged during this time would be more influential than the Education for All Handicapped Children Act (EAHCA) of 1973.

Serving Children with Disabilities and New Learners of English

The EAHCA began as a voluntary federal aid program that provided funds to state departments of education, school districts and universities. Although many states did provide some educational services for students with disabilities, many of these youngsters were either excluded from school or underserved relative to their educational needs. Coupled with the Vocational Rehabilitation Act of 1973, a civil rights law that prohibited discrimination in education on the basis of disability, EAHCA quickly moved to center stage in public education.

Within a decade of its enactment EAHCA and its subsequent reauthorizations helped to transform American public education. Nationwide about 11 percent of the school-age population today receives services through special education programs for students with disabilities. And almost 30 percent of all new dollars for public education in the later part of the twentieth century went to support expanded services to special needs students (Rothstein, 1995).

The 1970s also saw a surge in bilingual education programs for new learners of English. Successful litigation over issues of the segregation of language minority students, and inequity in the allocation of state and local funds between white and minority schools, led the courts to conclude that a unique curriculum, including native language instruction, was needed as a remedy. Cases like *Swann v. Charlotte-Mecklenburg Board of Education* in 1971; *Keyes v. School District Number One, Denver, Colorado* in 1973; and *Lau v. Nichols* in 1974 contributed to the desegregation of limited English speakers and the development of bilingual and bicultural programs in many communities across the nation (San Miguel, 1987).

A Tax Revolt

But the decades of the 1970s and 1980s were not only about growing educational programs and increased spending for public schools. An economic recession combined with inflation had many Americans looking for ways to cut back on expenses and they quickly turned to the schools. Most states still used local property taxes as the bulk of support for schools. As double-digit inflation continued year after year, the market value of a typical home also grew along with the assessed value of the home. The result was skyrocketing property taxes, and a tax revolt ensued.

In 1978, the voters in California, through a voter initiative, enacted Proposition 13, a move to limit property tax growth. This action inspired other similar efforts in states that allowed voter-initiated ballot questions and caused state legislatures across the country to scramble to shift school tax revenue away from a high dependence on local property tax to other sources.

Things shifted at the federal level as well. With the election of Ronald Reagan in 1980, many of the categorical programs of prior administrations were collapsed into federal block grants. While not successful in achieving Reagan's full education initiative (i.e., elimination of the U.S. Department of Education), block grants to states of all federal education funding, and school tuition vouchers for all funded in part with federal money, allowed him to accomplish much of his agenda.

A Nation at Risk

In an unanticipated turn of events, however, President Reagan also inspired the education reform movement that exists to this day. Under the leadership of his secretary of education, Terrel Bell, the Reagan administration launched the modern-day public school reform era characterized by accountability, standards and testing. The impact of the Reagan administration's report *A Nation at Risk* (National Commission on Excellence in Education, 1983) sent shock waves through the country that are still felt today. Education moved to the top of the domestic policy agenda. State and local funding for education increased once again, but this time in areas like testing, courses for advanced students and reduced class size. And Congress and succeeding administrations reacted by increasing appropriations to the very federal education programs President Reagan originally wished to eliminate.

The current era of school reform, now approaching its fourth decade, has been characterized by the development of academic standards at the state level and extensive accountability measures focused on student testing. The federal government played a central role in spurring and sustaining the movement. Through the leverage of federal education programs like the No Child Left Behind Act, states have been positioned to establish academic and performance standards, and to build state-wide testing programs to measure student academic achievement (Ravitch, 2010).

The financial impact of the ongoing school reform initiative has only been partially realized. Money spent to develop, administer and score tests has increased substantially. Estimates of the costs associated with the large-scale assessment movement range into the billions of dollars (Sacks, 1999). Many states and school districts have added resources to either create incentives for better school performance or to target funding to student populations struggling to meet established academic standards.

History Repeats?

School finance trends influenced by politics and school reform efforts like choice, site-based funding, incentive pay plans and litigation about the adequacy of funding are recycling through the system. Below are some examples.

Many policy makers are baffled by the nuances of school improvement and the challenges of educating at-risk populations of students. As a result, they have settled on a plan to dismantle or reorganize low-performing schools by various methods of private school vouchers for students, wresting control of schools from the school district by converting it to a charter school or privatizing the operation. Choice also serves the political purpose of diverting resources away from school districts with employee bargaining units, thus diluting the power of these employee groups to influence decisions about funding for schools.

Another trend that cycled through the education system was site-based management, which included providing greater discretion at the school level for the expenditure of financial resources. The theory at play is that school-based professionals have the best idea about the needs of their students and thus should be free to deploy available resources. Today they are called "innovation schools."

In other cases school boards or state legislatures are tampering with teacher and administrator compensation systems in an attempt to modify the behavior of school personnel. Merit pay, pay for performance and strategic compensation plans are examples. A hundred years ago the emphasis was on standardizing educator pay to attract and retain talented teachers and administrators.

But the trend most likely to have a major impact on school finance is litigation related to the adequacy of funding for all students in order that each student may achieve the established academic and performance standards. And more rural schools are now claiming the system shortchanges them most.

As trend data build in future years, the persistent problem of groups of students not meeting standards will likely lead to court challenges of funding systems that do not afford students the opportunity to reach established levels of learning. Historical patterns in American public education underscore the iterative nature of the development of the education system as it moves from crisis, to reform, to reaction, and on to a new crisis. School finance will remain a central aspect of the development of the public schools into the future.

Summary

The need for education and the value held for education has existed throughout history and throughout human cultures. The creation of the American experiment was predicated on the idea that average citizens could govern themselves. Inherent in this belief was the idea that education must be available to all.

A big challenge over time has been how to pay for universal education. This chapter traced the historical development of funding for schools from the colonial era to the present. One theme that emerges from the chapter is the belief in America that education is central to the national well-being and the willingness of Americans to support their schools. The struggle to build an education system and provide access to education for all Americans has been long and arduous. Questions of adequacy and fairness in school funding persist to this day.

Chapter 3 proffers questions about the relationship between money and student learning. It investigates notions of the relationship between money spent and the amount of learning students achieve. As we read in this chapter, the struggle to build and fund the education system we have today was a long and difficult one. Yet, some still ask, "is all this spending worth it?"

References

Adams, D. W. (1995). *Education for extinction.* Lawrence, KS: University Press of Kansas.

Alexander, K., and Alexander, M. D. (2004). *American public school law* (6th ed.). Belmont, CA: Wadsworth Group.

Anderson, J. D. (1988). *The education of Blacks in the south, 1860–1935.* Chapel Hill: The University of North Carolina Press.

Bailyn, B. (1960). *Education in the forming of American society: Needs and opportunities for study.* New York: Vintage Books.

Bolton, H. E. (1939). *Wider horizons of American history.* Notre Dame, IN: University of Notre Dame Press.

Brown v. Board of Education of Topeka, Kansas, 347 U.S. 483, 74 S. Ct 686. (1954).

Bullock, H. A. (1970). *History of Negro education in the South: From 1619 to the present.* New York: Praeger Publishers.

Columbus, C. (1987). *The log of Christopher Columbus.* (R. H. Fuson, Trans.). Camden, ME: International Marine Publishing. (Original work published 1493).

Cremin, L. A. (1970). *American education: The colonial experience 1607–1783.* New York: Harper & Row.

Cremin, L. A. (1980). *American education: The national experience 1783–1876.* New York: Harper & Row.

Cubberley, E. P. (1948). *The history of education: Educational practice and progress considered as a phase of the development and spread of western civilization.* Cambridge, MA: Houghton Mifflin Company.

Gonzalez, G. G. (1990). *Chicano education in the era of segregation.* Philadelphia: The Balch Institute Press.

Harrington, M. (1962). *The other America.* New York: Macmillan Publishing Company.

Johns, R. E., Morphet, E. L., and Alexander, K. (1983). *The economics and financing of education* (4th ed.). Englewood Cliffs, NJ: Prentice Hall.

Katz, W. (Ed.). (1969). *History of schools for the colored population.* New York: Arno Press and The New York Times.

Keyes v. School District No. 1, Denver, Colorado, 413 U.S. 189, 98 S. Ct. 2686. (1973).

Lau v. Nichols, 414 U.S. 563, 94 S. Ct. 786. (1974).

Mendez et al. v. Westminster School District of Orange County et al., 64 F. Supp. (San Diego, CA, 1946).

Moore, J. T. (1982). *Indian and Jesuit: A seventeenth-century encounter.* Chicago: Loyola University Press.

National Commission on Excellence in Education. (1983). *A nation at risk: The imperative for school reform.* Washington, D.C.: U.S. Department of Education.

Nieman, D. G. (Ed.). (1994). *African Americans and education in the south, 1865–1900.* New York: Garland Publishing, Inc.

Our documents (2012). Transript of Northwest Ordinance, 1788. Retrieved from www.ourdocuments.gov.

Peterson, P. E. (1985). *The politics of school reform 1870–1940.* Chicago: The University of Chicago Press.

Peterson, T. T. (1975). *Public school finance.* (Unpublished manuscript). University of Utah.

Plessy v. Ferguson, 163 U.S. 537, 16 S. Ct. 1138. (1896).

Power, E. J. (1970). *Main currents in the history of education.* New York: McGraw-Hill Book Co.

Prucha, F. P. (1979). *The churches and the Indian schools 1888–1912.* Lincoln: University of Nebraska Press.

Ramirez, A. (2009). State and government role in education. In E. Provenzo (ed.) *Encyclopedia of the social and cultural foundations of education.* Thousand Oaks, CA: Sage Publications.

Ravitch, D. (2010). *The death and life of the great American school system: How testing and choice are undermining education.* New York: Basic Books.

Roberts v. City of Boston, 59 Mass. (5 Cush.) 198 (1849).

Rothstein, R. (1995, Summer). Where has the money gone? *Rethinking Schools, 6.*

Sacks, P. (1999). *Standardized minds.* Cambridge, MA: Perseus Books.

San Miguel, G., Jr. (1987). *"Let all of them take heed": Mexican Americans and the campaign for educational equality in Texas, 1910–1981.* Austin: University of Texas Press.

Swann v. Charlotte-Mecklenburg Board of Education, 402 U.S. 1, 91 S. Ct. 1267. (1971).

Szasz, M. C. (1988). *Indian education in the American colonies, 1607–1783.* Albuquerque: University of New Mexico Press.

Tyack, D. (1974). *The one best system: A history of American urban education.* Cambridge, MA: Harvard University Press.

Tyack, D., James, T., and Benavot, A. (1987). *Law and the shaping of public education, 1785–1954.* Madison: University of Wisconsin Press.

Money and Learning

3

Aim of the Chapter

THIS CHAPTER EXPLORES THE DEBATE REGARDING THE role of money in schools and its relationship to student achievement. Several major studies are reviewed. Economic and educational perspectives are brought to bear on the deliberation. An effort is made to sort out the facts from the myths regarding the relationship between school spending and student achievement.

Introduction

Does money make a difference? Is the problem of poor student achievement due to a lack of financial resources? How is it that per pupil spending has increased substantially in school districts across the country during the past one hundred years (see Figure 3.1), yet some argue that student achievement has risen only slightly or remained flat? Why does the achievement gap between groups of students persist in so many schools, even in schools that spend more per pupil than their state average? But if the schools are doing so poorly, how is it that the United States remains the leading world economy, the leader in Nobel Prize awardees and the leader in technological innovation, and has the world's best higher education system?

There are many paradoxes associated with the question of money and schools. Various scholars and policy leaders stand on one side or the other of the issue. Some argue that a lack of financial resources keeps schools in a perpetual state of being underfunded. They see schools as barely being able to keep up with the annual rate of inflation, thus having less spending power each year. Other scholars and policy leaders assert that few schools can demonstrate a relationship between more funding and increased learning. New programs, new curriculum and endless national, state and local school reform initiatives produce no change. One side clamors for more money while the other side calls it a waste of resources.

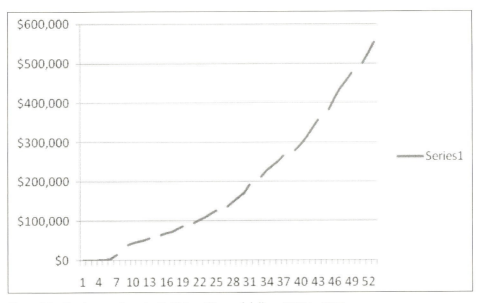

Figure 3.1   Total expenditure for K–12 in millions of dollars, 1900 to 2007.

The Economic View

The relationship between human capital and the economic well-being of individuals and nations has long been the focus of theorists from fields as wide ranging as economics, sociology and education (Smith, 1776; Hanushek and Kimko, 2000). Economists use the term "human capital" to describe a valuable resource, labor, used to produce wealth. Prior to the Industrial Revolution of the late eighteenth and early nineteenth centuries, labor combined with natural resources were the most significant determinants to a country's economic well-being.

Fertile lands could produce an abundance of crops, which could be exported for other trade items or cash. Good, deep harbors could facilitate trade and access to world markets. Supplies of iron and coal could be exploited for domestic development and export, etc. A vast supply of cheap, untrained and unskilled labor worked the fields, manned the ships and dug in the mines to support those economies.

The Industrial Revolution added a third element to the creation of wealth—physical capital—in the form of machines and industrial organization. In the past, for example, a family could work cooperatively to weave cloth for personal use and sell any surplus in the local market. The family might even have a contract with a merchant or trader to make extra yards of material.

But production under these circumstances was limited to what each family member could manufacture. The introduction of industrial machines to the manufacturing process vastly increased what one worker could produce. An industrialist would invest his own money, or entice others to invest, in the purchase or development of machinery designed for mass production. Workers were then hired to run the equipment and service the production process. Production was exponentially increased and trade greatly stimulated.

As in the past, this shift in economies also depended on a supply of plentiful cheap labor, but it soon became apparent that some level of education in the workforce was needed to support industrialization. Enhancement of the human capital in a country was required to sustain new and expanding industry. Mechanics were needed to keep machines operating, low-level managers were needed to oversee ever-expanding work groups, clerks were needed to keep track of vast amounts of raw material and finished products and more complicated machinery and manufacturing processes necessitated trainable workers. Thus, as countries increased industrialization, the demand for a workforce with a basic education also increased.

However, questions about the efficacy of investment in education, and who should pay for it, quickly emerged and continue to this day to shade the policy debates about funding for education, access to education and the return on investment (Grubb, 2009). Should industries be taxed to support a system of schooling to develop a workforce with a basic education? Is education a private matter because better-educated individuals tend to earn more and thus benefit more directly from investment in education? Should parents be responsible for their children's education and each individual adult responsible for his or her continuing education?

On the other hand, does society as a whole reap some benefit from an educated population, and so society should support education? Is there a best mix of publicly and privately funded education? Such debates about who should pay for education and how much should be spent occur at the local, state, national and international levels (Checchi, 2006). They have for generations and likely will continue.

Individual families also grapple with questions about return on investment as they chart the course of education for their own children (Altonji and Dunn, 1996). Should the toddlers be sent to preschool? Should it be a public or private preschool? Is a private school K–12 education a good investment? What about the new local charter school as an option? Should the family relocate to a community with a reputation for better public schools? Does the child have a unique genius that should be nurtured? Will this special talent grow on its own or does it need special schools and teachers? What is best for a child with special education needs? Should the children be encouraged to apply to selective private colleges, compete for seats at the state's flagship public universities, go to a public regional college or stay at home and attend the local community college for two years? Each individual family negotiates this decision making territory.

While today's global economy still rests on traditional market forces like labor, physical capital and natural resources, data from international comparisons demonstrate the clear relationship between the educational attainment achieved by a nation's citizenry and its economic standing. Nations with higher overall levels of education have stronger economies (Organization for Economic Cooperation and Development [OECD], 2011). Furthermore, international competitiveness, now more than ever, lies in the human capital of each nation and may be the most significant contributor to a nation's gross domestic product (GDP), the sum of all goods and services produced by a country. In fact, estimates for the return on investment in education, as measured by GDP, exceed those to be found from investment in natural resources or physical capital (McMahon, 1991).

Consider the multitude of underdeveloped and developing nations with abundant natural resources, large populations of poorly educated citizens and devastated economies where great disparities in personal income exist among their citizens. Such nations are characterized by a wealth distribution within the population that has a vast number of poor and destitute families, a small middle class and a very small, super-rich group who benefits from exploiting the country's natural resources. Democratic institutions are typically scarce in such nations while authoritarian governments are common.

On the other hand, think of countries like Japan, with few natural resources, that maintain viable national economies and have a thriving middle class and very little poverty. Developed nations have come to appreciate the economic and social value of a comprehensive system of education, even as they struggle to craft education policies that will achieve the best return on investment. Similarly, families recognize the value of education and make choices about how much to invest in their children's education (Becker and Tomes, 1986).

Beyond the economic returns on investment in education, nations receive social benefits in proportion to the relative level of education in the society. Such non-market benefits are manifested in a range of improved social conditions from reduced crime, to lower welfare needs, to better health, to more charitable giving, to more participation in civic life (Wolfe and Zuvekas, 1997). All of these social benefits enhance the quality of life within a nation and relieve the need for taxes and private moneys that would otherwise have to be spent in these areas.

For example, a reduced crime rate lowers the tax burden on society that would have to go to pay for more police, courts and prisons. Private investment also benefits from reduced crime in the form of lower personal protection needs, lower insurance rates and lower taxes for crime deterrence. A similar case can be made in the areas of healthcare and welfare. Public and personal money not spent in these areas can be used for economic investment, investment in a nation's infrastructure, personal improvement, for other public purposes to improve the national economy or to enhance personal wealth.

Additionally, a better-educated populous in a dynamic economy will also come to value liberty as an essential element for personal and national well-being. Democratic institutions flourish within better-educated populations. Civic engagement becomes important as a personal interest and a societal obligation. Thus, increased education across the society has an upward spiraling effect for the nation.

The United States is a good case in point. As you read in chapter 2, from the beginning of the colonial period education was viewed by many leaders in America as essential to economic development and the formation of civic society. Many of the founders of the republic proffered ideas for universal education as indispensable to the creation of the new democratic republic, the preservation of liberty and the rejection of tyrants.

Picture 3.1   Education is costly, but the benefits to the individual and society are great.

For example, John Adams, who would go on to be the second president of the United States, drafted eloquent language about education for the original Massachusetts constitution. That constitutional language explicitly outlined the need for education as essential to the promotion of science, industry, the arts and democracy. It also went on to commit the state in assuming the responsibility to promote education for all (McCullough, 2001). Adams made clear that the advancement of society was tied to the broad distribution of education within the populous to the benefit of the individual, the economy and the greater democratic society.

Personal Choice and Systemic Failure

Questions of access to education, the quality of education and equity within the educational system interplay among personal choices and public policy targeted toward greater societal gain from education (Filmer and Pritchett, 1999). As do families, nations also ruminate about the proper balance between investments in education and return (Hanushek, 1986). Yet despite the array of benefits that accrue to individuals and society, resistance to educational opportunities by individuals and the failure of extant education systems to educate large numbers of certain students to higher levels persists.

This is a particular problem in the United States. In the parlance of the contemporary literature this is called the "achievement gap" (Cameron and Heckman, 2001; Carpenter, Ramirez and Severn, 2006; Carpenter and Ramirez, 2007). Throughout the history of public education in America, some groups, on average, have not reached established levels of learning. The system has failed to produce the individual and societal benefit it was designed to deliver.

Explanations of these achievement gaps range from the innate ability of the student groups, to ineffective schools, to bad family background, to lack of community support, to invidious systemic malevolence. Scientific studies and court battles have delved into the causes of and solutions to the achievement gaps. Accusations and blame abound as education and policy leaders assume personal perspectives on the issue and fall back on arguments about an inequitable system that treats different groups differently and favors some students over others. Others take the side that the system is incapable of fixing problems that emanate from outside the school. Money inevitably moves to the center of the debate.

More importantly, the larger question of persistence in education and particularly intergenerational educational attainment emerges as a critical policy issue. Shrinking the achievement gaps is a national priority for personal and societal reasons in the United States. Educated individuals tend to achieve a better overall lifestyle, and they contribute more to society in the form of taxes, charitable giving and civic engagement. They require less from society in the form of welfare or the criminal justice system. An undereducated subpopulation that does not fulfill its potential for itself or the greater society can end up being a net loss to the nation (Psacharopoulus, 1994; Levin, 2009). Nations that cannot develop sufficient

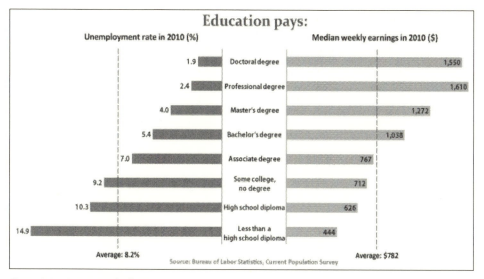

Figure 3.2   Earning and education.
Note: Data are 2010 annual averages for persons age twenty-five and over. Earnings are for full-time wage and salary workers.
Source: Bureau of Labor Statistics, Current Population Survey.

high-skilled human capital within their native population must rely on immigration to meet economic demands. Within the global economy, this growing need for highly educated human capital is becoming ever more competitive (Checchi, 2006).

The global economy and global economic competition put a premium on a country's human capital. The education system of a nation is integral to developing human capital and competing well economically in the global marketplace. How much national treasure should be invested in education is a persistent question. Developed nations like the United States, Japan and Germany have a fraction of the population of developing nations like India and China. As these developing countries invest more in their education systems, their stock of human capital will increase immensely.

Therefore, a new concern emerges among developed countries that is often called "the underachievement gap." This gap is defined as the shortfall between predicted educational attainment compared to actual educational attainment, particularly for subpopulations with full access to education and supporting resources. Many questions arise from this issue: why levels of secondary school education are not higher in the population as a whole after accounting for poverty and other conditions; why the percentage of postsecondary enrollments is so low; why the success rate of degree-seeking students is so low; and why is there a population of youth who regresses from intergenerational educational attainment.

Current research on "the achievement gap" reveals that claims of a monolithic gap between minority and majority student populations in the United States were an incorrect conceptualization of the problem (Carpenter et al., 2006; Carpenter and Ramirez, 2007). This research discovered varied and nuanced "gaps" between and within the groups studied, i.e., black, Hispanic and white. Furthermore, the research showed that variables that predicted poor student achievement for the various groups did not predict dropout behavior. Importantly, race and ethnicity were not predictor variables in either study. Thus, policy initiatives that are recommended to target the poor academic achievement of these groups should be based on a deeper understanding of local conditions and not broad generalities.

Questions related to why education systems fail to deliver higher numbers of completers at all levels of education are central to contemporary research in America. The elusive question of the underachievement gap is of vital importance. Can the underachievement gap be defined, quantified and explained? What policy initiatives should emanate from this new research? The importance of this challenge to the national education and economic policy is significant.

Demographic conditions in the United States along with the international competition for human talent make the possibility of reaching greater efficiencies within the U.S. education system a paramount goal. If the United Sates cannot get better results from its system of elementary, secondary and tertiary education, future prospects for continued economic prosperity and social cohesion are in jeopardy. One group that offers a promising prospect for improvement is comprised of those individuals who have the means, access and ability to reach higher levels of education—but have not. Understanding this underachievement gap will lead to policy initiatives that will mitigate the loss of and enhance human capital within the nation. Changes to this calculus will support the economic viability of the United States. Combined with efforts to mitigate the achievement gaps among the poor and some minority groups, a major increase in the overall level of education within the American population can be accomplished. Will more money be needed to accomplish this?

Measuring Inputs and Outputs

The research and literature on money and schools is highly concerned with efficiency. Scholars strive to find the precise formula that will measure the inputs to the school system and then measure the outputs achieved with those resources. Economists and researchers refer to this as the production function (Krueger, 1999). Many of these scholars have been frustrated in their efforts. Their findings have been confounding, and by and large ignored, by educators and policy makers. There are many reasons for this. One of the biggest challenges to the study of efficiency in spending for schools is how to measure the variables involved.

Consider the question of teacher quality. Common sense says teachers have to be a major influence on a student's learning. Few would argue against the notion that a "good teacher" should get better results than a "bad teacher." But what is a good teacher? Is it qualifications? How should qualifications be measured? Is it the teacher's level of education? Are teachers with master's degrees more effective than those with bachelor's degrees? Is it the teacher's licensing status from the state? Do teachers with a state license perform better than those without a state license? What about teachers in private schools where state licenses are not required? Is it the teacher's commitment to students? How can commitment be measured? Could it be the teacher's ability to solve problems that inhibit student learning in the classroom? How can that be assessed (Kennedy, 2008)?

Surely class size is a significant input. A class with forty students cannot achieve the same level of learning as a class of twenty students, right? Does the makeup of the students in each group matter? How about the experience, motivation and ability of the teacher? Will a "bad teacher" get better results with twenty students rather than forty students? How small does a class have to be before the overall level of student learning in the class increases? How large a class is too large? At what point is student learning adversely affected? Can class size alone be a contributing factor to student achievement?

If the critical inputs are not teachers or class size, then it must be the curriculum that makes the difference. So a simple survey of schools across the nation should reveal which schools achieve the highest levels of learning and what curriculum they use. But strong schools and weak schools often use similar curricula. Educational philosophies and curriculum based on perennialism, scholasticism, progressivism or postmodernism have all been shown to fail and succeed. Similarly, models of teaching founded on direct instruction and exploratory learning are promoted with equal enthusiasm by their respective proponents as the best approach. How can it be that very diverse methods of teaching or curriculum can be effective in one place and not another?

Perhaps it is a school's organization that is making the difference. We see kindergarten through fifth-grade schools, and K–8 schools, and middle schools comprised of grades six, seven and eight, and nine through twelve high schools, and ten through twelve high schools, and freshmen/ninth-grade centers, and schools within schools, and every manner of school organization. In the middle of the twentieth century education leaders and policy makers touted the value and need for the large comprehensive high school. By the end of the century education innovators were advocating for "small" high schools as the solution. Sometimes these organizational structures seemed to help and sometimes they did not.

It is also argued that the biggest and most significant input is the student. A student's family circumstances can either support or diminish a student's success in school. Family poverty has historically and frequently been cited by researchers as being associated with low educational performance (Coleman, 1966). Issues of race, ethnicity, social class and home language have all been investigated separately and in combination in a search for answers about why we have so many poor-performing and underperforming students. This underperformance persists despite the ever-increasing sums of money put into the public and private education systems.

But economists, education researchers and policy makers also understand that correlation is not causation. While they can easily link variables like poverty and student achievement, it is much harder to explain why the link exists and even harder to determine a means of improving the situation. Policy makers and researchers are limited by what can be measured given the existing state of knowledge and the capability to analyze available data.

On the other hand, educators and school leaders have a more practical view of the system. They understand that reducing resources for the education system cannot make things better. From the school leader and classroom perspective the needs are great. Students who are behind need extra support and students who can excel should not have their potential inhibited. Most pre-collegiate educators support the idea that the aim of the system is to prepare students for success beyond high school. This means a student who will be a lifelong learner able to continue his or her education in a postsecondary institution or on the job, be an engaged citizen, and establish a good quality of life. Educators understand that an inadequately prepared student is most likely sentenced to a life of poverty.

We can measure the inputs to the system and measure the outcomes of the system. The challenge remains in linking the two in a causal relationship. Thus, critics remain entrenched in their beliefs about wasted money, and advocates are left with weak rationales to justify requests for more money. Somewhere between spending no public money on education and spending vast amounts more is the answer.

Where Did the Money Go?

No matter how one calculates the total, the amount of money spent on pre-K–12 education has expanded substantially since the beginnings of the public schools. These increases cannot be attributed to inflation alone. In the nineteenth and early twentieth centuries the increase in spending was devoted to establishing a system of schools, building the system's capacity to fulfill its mission and serving a student population that expanded by leaps and bounds. Throughout that history critics argued against increased spending as unnecessary and wasteful. Considering that history in retrospect, would America be the nation it is today without having built and paid for its system of schools? Or would America have developed like other "new world" nations that chose not to commit to and fund a comprehensive system of education?

During each epoch, policy leaders have made decisions to spend more on education based on their judgment about the best interests of the nation. In the modern era, new critics also argue that increased spending is not needed and wasteful. They claim increased spending is like trying to fill a bottomless pit with dollars and point to stagnant test scores as their evidence. However, an analysis of the facts reveals a more complex answer to questions of spending and where the money went. Today America is still building the capacity of the system to fulfill its mission and serve an expanding student population.

Since the 1950s the nation has struggled to dismantle a dual system of schools— one affluent and white, and one poor and minority. Battles over desegregation and the equitable distribution of funding persist to this day. Desegregation was a moral and legal necessity that mandated there must be one system of schools in each state. Litigation over school funding resulted in the determination that all children in a state will have an equal educational opportunity. Correcting these constitutional infringements has been costly and necessary. Another issue from the 1950s was the abysmal graduation rate for the nation as a whole. During that time only about half of the youngsters in the public schools went on to graduation. This statistic was much worse for minority groups. This poor completion rate was seen as a national problem and efforts were made to increase the number of high school graduates.

At various times policy makers have determined that expanding services to a broader population of students is in the interest of society. For example, they determined in the late nineteenth century that a sixth-grade education was not enough and that high schools would be publicly funded. Today we see services extended to full-day kindergarten, preschools, and alternative education for dropouts. The government even subsidizes higher education through its aid to colleges and universities, and by underwriting grants and student loan programs. This is done because policy makers believe our nation will benefit.

Reduced class size has been pursued by educators and policy makers because it is believed more individual attention for each child will help all students. Class size reduction is a costly policy because it not only requires that more teachers be added to the payroll, but also necessitates additional capital outlays to pay for more classrooms. This policy has been undertaken despite little empirical evidence that student learning will increase (Krueger, 1999; Hoxby, 2002). However, common sense informs parents, teachers and some policy makers that smaller class size is a worthwhile expense.

Additional personnel with various roles have been added to the payroll of schools because policy leaders believe certain services and support outside the classroom are important. Over the past several decades new positions have been created in the public schools for jobs that didn't even exist three decades ago. Computer network specialists and trainers are examples. Computer education and the use of similar technology have added enormous costs to the operation of schools for hardware, software and specialized personnel. These technology resources and expanded curriculum were added to the schools despite the fact that not one research report existed to show learning would increase. Digital technology is now a ubiquitous aspect of the schools. Who would argue that it should be pulled out because test scores have remained flat?

Since the 1970s special education services have greatly expanded within the public schools. Prior to this time many children with disabilities were excluded from the schools or admitted on a voluntary basis. When the federal government adopted the Rehabilitation Act of 1973, it established civil rights legislation regarding the handicapped in almost all aspects of our society, including the schools.

This policy ushered in a major shift in America. For schools it meant an enormous adjustment in order to build, in essence, an entire new system within the PK–12 system. Richard Rothstein (1995) estimates that about thirty cents of every new dollar added to the schools went for building the special education system. Even the federal government knew it would cost a lot more to add special education to the PK–12 system and contemplated contributing up to 40 percent of the excess cost of the then-new initiative. Of course they never contributed more than 14 percent, and so states and local school districts have been shouldering the biggest part of the expense.

This new population of students with disabilities precipitated the hiring of more teachers and other professionals. Specialists of all kinds are now standard in the schools across the country in order to diagnose various disabling conditions and prescribe the array of associated therapies and teaching strategies. In addition numerous new professionals and support personnel have been added to the payroll to facilitate the educational experience of children with disabilities. America chose to establish this right for children with disabilities, despite the fact that test scores were not likely to rise. It was done because America and its policy makers believed it was the right thing to do.

Other less dramatic, although very expensive, events have also affected the cost of schooling while not contributing to improved test scores. Millions of dollars were spent on the asbestos abatement programs of the 1970s, 1980s and 1990s. This building material, asbestos, was once considered a harmless and useful building product. Many schools were built using ample amounts of asbestos because it was a cheap and versatile building material. When science determined that asbestos was a carcinogen, programs were initiated to remove it from schools. No sane person will argue that children and school personnel should have to attend school or work in unhealthy environments because of money concerns.

Rising energy costs have plagued schools just as they have all aspects of American society. In the 1960s school buses were able to chug around on fuel that cost twenty-five cents a gallon. Today the same gallon of fuel costs 2,000 percent more! Similarly, the cost of heating, cooling and lighting schools has gone up. Remember, too, as outlined above, schools have also added greatly to energy consumption with the addition of all the new technology.

An even more mundane yet very expensive part of operating schools is insurance. Schools, like any other major enterprise, must buy large amounts of insurance to cover the risk associated with fire, natural disasters, vandalism, liability, transportation and the errors and omissions of their employees and board members. Additionally, school districts compensate employees by giving them a benefit in the form of health insurance. Insurance rates of all kinds have been increasing by leaps and bounds since the 1980s with no end in sight.

Teacher salaries represent a big part of a school district's budget, and personnel costs overall constitute between 65 percent and 85 percent of what school districts spend on average each year. In America, employers use pay and benefits to attract and retain employees. Intelligent, educated, qualified and motivated personnel are highly sought after by employers and thus cost money. Current education policy dictates that every student have "highly qualified" teachers. The days of teachers being hired irrespective of qualifications, because of political patronage or connections, are mostly gone.

Text Box 3.1 Simple models have difficulty explaining school success and failure.

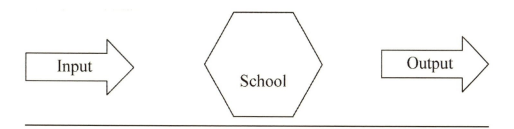

Of course, schools spend money in rational ways to reach the goal of increased student learning; that is their mission. Yet it is often the case that financial resources are spent in ways that do not directly support this goal. Circumstances beyond the control of the school, as in the case of asbestos abatement or public policy that is perceived to be in the broader interest of the society, necessitate spending that is removed from the education mission. Decisions about money and spending exist in a broader context than test score results.

Summary

Money does make a difference in schools. An optimal amount of resources is tied to the ability of a school to fulfill its role in society. Americans have historically committed to funding schools because of the belief that it is essential to the viability of the nation. The nation's prosperity is directly linked to the overall level of education within the population. Investment in education is of national importance. Yet the matter of societal versus personal benefit from education fuels the controversy over who should pay for schools.

Issues of failure within the system to educate large numbers of students persist. In addition, the lost potential of students who comprise the underachievement gap is emerging as a national concern. The controversy over the inability of researchers, policy makers and educators to demonstrate clear causal relationships between funding and educational results invigorates the debate over money and schooling. Much of school spending does not directly tie to academic achievement, but policy makers have determined such spending to be essential.

Money for education is an elusive matter. Determining how much is needed and how much is sufficient is not a precise science. Perspectives not only differ about how much should be spent but also where to spend it. Claims and counter-claims about need and waste abound. The research literature on the issue of money and schools is extensive but offers little definitive information.

School critics decry what they say is excess and waste within the education system. They say money doesn't matter, anyway, because no clear relationship exists between spending and student achievement. But yet, these same critics are often silenced when the issue is brought down from the abstract to the personal. Perhaps what they are saying is that "money doesn't matter for your child, but it certainly matters for mine."

In chapter 4 the discussion about school finance moves to the courts, and questions of money for schools, how much and for whom take on legal dimensions that revolve around constitutional issues. Much of the same economic and educational research brought out in this chapter is viewed through the jurisprudence lens. Legal argument tends to put a sharp focus on some of the less clear school finance issues considered thus far in earlier chapters.

References

Allegretto, S., Corcoran, S., and Mishel, L. (2008). *The teaching penalty: Teacher pay losing ground*. Washington, DC: The Economic Policy Institute.

Altonji, J., and Dunn, T. (1996). The effects of family characteristics on the returns to schooling. *Review of Economics and Statistics, 78*(4), 692–704.

Becker, G., and Tomes, N. (1986). Human capital and the rise and fall of families. *Journal of Labor Economics, 87*(6), 1153–1189.

Cameron, S., and Heckman, J. (2001). The dynamics of educational attainment of black, Hispanic and white males. *Journal of Political Economy, 109*(3), 455–99.

Carpenter, D., and Ramirez, A. (2007). More than one gap: Dropout rate gaps between and among black, Hispanic, and white students. *Journal of Advanced Academics, 19*(1), 32–64.

Carpenter, D., Ramirez, A., and Severn, L. (2006). Gap or gaps: Challenging the singular definition of the achievement gap. *Education and Urban Society, 39*(1), 113–127.

Checchi, D. (2006). *The economics of education: Human capital, family background and inequality*. Cambridge, UK: Cambridge University Press.

Coleman, J. (1966). *Equality of educational opportunity*. Retrieved from http://eric.ed.gov/ERICWebPortal/search/detailmini.jsp?_nfpb=true&_&ERICExtSearch_SearchValue_0=ED012275&ERICExtSearch_SearchType_0=no&accno=ED012275.

Filmer, D., and Pritchett, L. (1999). The effects of household income wealth on educational attainment: Evidence from 35 countries. *Population and Development, 25*(1), 85–120.

Grubb, W. (2009). *The money myth: School resources, outcomes and equity*. New York, NY: Russell Sage Foundation.

Hanushek, E. (1986). The economics of schooling: Production and efficiency in public schools. *The Journal of Economic Literature, 24*(3), 1141–77.

Hanushek, E., and Kimko, D. (2000). Schooling, labor-force quality, and the growth of nations. *American Economics Review, 90*(5), 1184–208.

Hedges, L., Laine, R., and Greenwald, B. (April 1994). Does money matter? A meta-analysis of studies of the effects of differential school inputs on student outcomes. *Educational Researcher, 23*, 5–14.

Hoxby, C. (2002). The effects of class size on student achievement: New evidence from population variation. *Quarterly Journal of Economics, 115*(4), 1239–85.

Kennedy, M. (2008). Sorting out teacher quality. *Phi Delta Kappan, 90*(1), 59–63.

Krueger, A. (1999). Experimental estimates of education production functions. *Quarterly Journal of Economics, 114*(2), 497–532.

Levin, H. M. (January 2009). The economic payoff to investing in education justice. *Educational Researcher, (38)*, 5–20.

McCullough, D. (2001). *John Adams.* New York, NY: Simon and Schuster.

McMahon, W. W. (1991). Relative returns to human and physical capital in the U.S. and efficient investment strategies. *The Economics of Education Review, 10*(4), 407–37.

Meyer, R. H. (1996). Value-added indicators of school performance. In Hanushek, E. A., and Jorgenson, D. W. (Eds.) *Improving the performance of America's schools* (pp. 197–223). Washington, DC: National Academy Press.

Psacharopoulus, G. (1994). Returns to investment in education: A global update. *World Development, (22)*9, 1325–43.

Organization for Economic Cooperation and Development (2011). *Education at a glance 2011: OECD indicators.* Paris: Organization for Economic Cooperation and Development. Retrieved from http://www.oecd.org/dataoecd/61/2/48631582.pdf.

Rehabilitation Act of 1973. Public Law 93-112 93rd Congress, H. R. 8070. September 26, 1973.

Rothstein, R. (1995). Where has the money gone? *Rethinking Schools, 6.*

Smith, A. (1776). The wealth of nations. In Cannan, E. (Ed.), *An inquiry into the nature and causes of the wealth of nations* (5th ed.) (1904). London, UK: Methuen & Co. Retrieved from www.econlib.org/library/smith/smwn.html.

Wolfe, B., and Zuvekas, S. (1997). Non-market effects of education. *International Journal of Education Research, 27*(6), 491–502.

School Finance and the Courts 4

Aim of the Chapter

IN THIS CHAPTER THE RELATIONSHIP BETWEEN THE courts and school funding policy is explored. The reader will gain an understanding of how school funding schemes are affected by constitutional protections afforded each individual, and how constitutional issues and court decisions interrelate with state school finance systems. Selected court cases are discussed to illustrate key legal turning points that have influenced policy and thinking about school funding.

Introduction

The courts have played a pivotal role in defining the legal boundaries within which state school finance systems exist. Through an iterative process of litigation, legal thinking about education finance has developed over time to its present-day status. These court interpretations of the law, conducted in forty-five of the states (National Access Network, 2007), often reflect changing societal values and norms. This legal thinking is applied to the allocation of resources for schools and the children who benefit from it. In some cases watershed decisions shift the course of education funding, while in other cases the status quo is preserved.

Sometimes, even when the plaintiffs lose, their cases influence the political processes of school funding. This is because the plaintiffs often present a strong, rational argument, which while not meeting an esoteric legal standard, persuades policy makers. The issue in contention might gain the support of the public, who in turn pressure policy makers for change. And, most commonly, the plaintiffs might have come close to a victory after mounting a strong attack, and the threat of the judiciary dictating policy spurs the legislature to act, thus maintaining control of their policy prerogatives. Finally, litigation, for all the reasons listed, can sometimes have the effect of breaking apart political logjams that heretofore had impeded legislative action to fix school funding laws. This can give political cover for legislators

---

Text Box 4.1 Reluctant defendants.

It is not unusual for defendants in school finance cases to be sympathetic to the cause of the plaintiffs. For example, in the school finance case *Committee for Educational Rights v. Edgar* (641 NE 2d 602, 267 Ill. App. 3d 18, 204 Ill), the state superintendent of education for Illinois at the time, Robert Leininger, was asked by a reporter how he felt about being named a defendant in the case. "I plead guilty," he replied, which set off a blizzard of letters and calls from the Illinois attorney general's office, legal counsel for the state.

Leininger had been a longtime advocate for fixing the state system for funding schools. He had addressed the issue of inequity in funding in many forums, including the state legislature. He was a true champion and major force for changing a system that allowed the highest-spending school districts to spend 500 percent more than the lowest-spending school districts. At the heart of the matter was overreliance on local property taxes.

The named defendant in the case, Governor Jim Edgar, showed great political courage after the state supreme court upheld the state's funding system. He publicly stated that the system was unfair and initiated a program to better equalize funding for schools. The price tag for the fix at that time was a billion dollars. As a Republican governor with major support from areas that included wealthy school districts, he did not have to take on the issue—no one expected him to do it. He did it because he knew it was the right thing to do.

---

and enable them to gain positions previously unavailable to them. The judiciary is prominent in shaping school finance policy.

The Role of the Courts

Our legal system presupposes a hierarchy within the law. At the top is the federal constitution, which begins by articulating those individual liberties and rights so cherished in our culture, and then defines the separate powers, responsibilities and limitations of the three branches of government and the individual states. Each state, in turn, adopts a state constitution as part of organizing its government. Among the powers of the legislative branch is the crafting of statutes, or laws, which serve as the operational guide for government and society. The executive branch is charged with implementing the law, while the courts function to interpret law.

As the courts make judgments about adopted laws and rules, in situations where no written law exists, it adds to the understanding and interpretation of the law through its decisions. This court-generated law, known as "case law" or "common law," adds to a large body of legal principles upon which future decisions are made. These legal principles are "precedents" that guide subsequent court decisions. Thus, applying these principles of law, interpreting laws adopted by the legislature and determining whether laws are constitutional are the main functions of the court. The question about the courts' authority to review legislative action was settled early in U.S. history. Legal scholars point to *Marbury v. Madison,* the case resolved in 1803, as establishing the power of the court to decide these matters.

Court systems, both federal and state, are segmented to provide a forum for initiating disputes, a mechanism for appeal and a court of last resort for final arbitration. For the purpose of this chapter they will be referred to as district court, court of appeals and a supreme court, state or U.S. It is not the role of the courts to encourage or solicit cases. Cases are brought before the court by aggrieved parties, upon which the court must determine whether it has jurisdiction in the matter, for example, a state or federal question. Courts do not issue advisory opinions, hear cases prematurely, hear from individuals who are not affected by the argument (i.e., have no "standing"), deal with hypothetical questions or consider issues that are no longer viable, that is, "moot" (Alexander and Alexander, 2009).

The U.S. Supreme Court hears school law cases for questions related to the constitutionality of a law or government action. These questions of constitutionality are concerned with the U.S. Constitution. Many school law cases do not rise to this level and are heard in state courts for questions of state law or state constitutional matters. Cases are selected for consideration by the highest court, state or federal, based on the same criteria listed above, often when lower courts have rendered contradictory opinions.

Picture 4.1   Motto over U.S. Supreme Court, "Equal Justice Under Law."

Federal Constitution

School finance questions that fall within the scope of the federal courts mostly revolve around a question of law related to federal legislation or the U.S. Constitution. This scope of jurisdiction will generally come under two broad areas: individual rights guaranteed under the U.S. Constitution or the structural provision of the constitution, which outline specific powers and duties of the federal government, the states and the people.

Federal education law—for example, a controversy regarding Title I of the No Child Left Behind Act—would also be disputed in federal court. Many school finance disputes do not meet these criteria for consideration in federal court. Fewer still focus on legal questions that will need the attention of the U.S. Supreme Court and are decided at the federal district or appeal level. However, those cases decided by the U.S. Supreme Court become "the law of the land" and affect the entire education system of the nation.

State Constitution

In the overwhelming number of school finance cases, it is the state constitution that is the battleground. From the earliest times when states drafted their state constitutions and set out provisions for education, litigations soon followed over the meaning and intent of the constitutional language. This perspective about the meaning of state constitutional language persists to this day.

Eastman (2007) presents a compelling argument regarding the nature and intent of the constitutional language found in the various state constitutions today. He points out how the meaning and intent of the language has been interpreted and reinterpreted by succeeding generations.

Each state has language in its constitution regarding public education. As such, the nature and interpretation of the language has broad implications for school districts and the students they serve. Phrases like "thorough and uniform system of education;" "a system of free common schools;" "the legislature shall encourage by all suitable means the promotion of intellectual, scientific, moral and agricultural improvement;" or "it is, therefore, the paramount duty of the state to make adequate provisions for the education of all children" appear in the constitutions adopted by the states. For some states these words trace back to a constitution first ratified by the state during the revolutionary era or upon the occasion of its admission to the Union. For other states it is part of a more contemporary constitution. What is clear from a review of state constitutional language is that the education of children and youth is of central importance.

Much of school finance litigation turns on the interpretation of the education provision in a state's constitution. Words like "adequate," "quality," "uniform," "thorough," and "system" are parsed and argued over. The intent of the drafters of the constitution is divined by lawyers and judges. In some cases the words are centuries old. Ultimately, the courts make judgments about the meaning and intent of the education articles in the state constitution.

Sometimes the court decides the words have literal meaning, and in other cases it is determined that the words are figurative. The courts have also ruled that the constitutional language is fixed in time and cannot apply to today, while in other states their courts conclude that the language must be interpreted within a contemporary context. Many of these decisions have enormous consequences for states, their political leaders, school districts, school administrators, teachers, students, taxpayers and communities. In some states billions of dollars are at stake.

Text Box 4.2 Litigations challenging constitutionality of K–12 funding.

In Process* (27)	No Current Lawsuit (18)	Never Had a Lawsuit (5)
Alaska	Alabama	Delaware
Arizona	Arkansas	Hawaii
Colorado	California	Mississippi
Connecticut	Florida	Nevada
Georgia	Idaho	Utah
Indiana	Illinois	
Maryland	Iowa	
Missouri	Kansas	
Montana	Kentucky	
New Hampshire	Louisiana	
New Jersey	Maine	
New Mexico	Massachusetts	
North Carolina	Michigan	
Oklahoma	Minnesota	
Oregon	Nebraska	
Pennsylvania	New York	
Rhode Island	North Dakota	
South Carolina	Ohio	
South Dakota		
Tennessee		
Texas		
Vermont		
Virginia		
Washington		
West Virginia		
Wisconsin		
Wyoming		

* "In Process" ranges from recently filed cases to cases where full implementation of the remedy seems close at hand.

Source: Hunter, M. A., National Access Network, 2007. Retrieved from http://www.school funding.info/litigation/In-progress%20 litigation.pdf.

Legal Battles Affecting School Finance

Presented below is a summary of some of the more significant cases that have in-fluenced legal thinking about funding schools. The subjects of some of the cases do not even consider public education or finance directly, yet they serve as a building block for later legal arguments that do address school funding. Cases have been grouped by categories for convenience, recognizing that simple classifications are often difficult to achieve in complicated legal matters. Furthermore, the list is by no means exhaustive, but rather intended to serve as examples, illustrative of cases that over time have contributed to the broader themes of education finance.

Education and the Constitution

It may seem odd that a discussion of school finance should begin with a case about school segregation. But the case of little Sarah Roberts (*Roberts v. The City of Boston*), the African American child who sought to attend an elementary school in her neighborhood rather than walk across town to the school established for black children, speaks volumes about how schools and their funding were viewed historically. Since the 1820s, Boston had established a separate elementary school for black children, and by the time Sarah started school, a second one existed. The African American population of Boston was barely 2 percent at the time, yet the struggle for equality was no less intense.

The idea that "separate but equal" was an acceptable legal doctrine was the decision of the court. The court even noted that, "the teachers of the school have the same compensation and qualifications as in other like schools in the city" as part of the legal rationale. The Boston School Committee did put a little more money into the black schools to repair facilities and furniture, and upgrade materials. The idea that equity in education could be achieved while preserving two systems, one white and one black, was firmly entrenched.

Minorities in other states also faced similar circumstances throughout the nine-teenth and early twentieth centuries, when schools were available to them at all, for example, Asians in California, Hispanics in the Southwest, and Native Americans throughout the nation. In some communities it was common for the white im-migrants to be segregated based on the language of their homeland. The idea that unique education delivery systems should be maintained based on race or some other distinction was part of the American zeitgeist.

It would take more than one hundred years before this legal concept of separate but equal would change. Even the Civil War amendments to the U.S. Constitution could not alter the pattern. The Fourteenth Amendment, adopted in 1868, with its exalted language about due process of law and the equal protection of the law for all citizens, was essentially impotent as state and federal legislatures and courts circumvented its purpose. In fact, it was the U.S. Supreme Court that enshrined the doctrine of "separate but equal" as the law of the land in its 1896 decision in *Plessy v. Ferguson*.

While not a school case, but one about public transportation, the legal principle applied. In the eyes of the court it was not illegal for a state to pass laws ordering the segregation of citizens based on race or other classifications. Only one dissenting justice, John M. Harlan, understood that our constitution was designed to serve just one class of person—a U.S. citizen—and other distinctions had no meaning before the law.

Sixty years later the doctrine of separate but equal would be rejected as not viable when another black child, Linda Brown, succeeded in her bid to desegregate schools in Topeka, Kansas. In 1954 *Brown v. Board of Education of Topeka, Kansas* was the watershed legal decision of the U.S. Supreme Court that ended the separate but equal doctrine. The words of the court rippled through American society. The court acknowledged that efforts had been made to equalize facilities, curriculum and personnel but decided that the Fourteenth Amendment prohibited the separate treatment by the government since it was inherently unequal. This would prove to be a huge factor in subsequent school finance litigation, as plaintiffs would claim this same constitutional protection.

The central importance of education to the individual and society was articulated in *Brown* as one of the reasons for the court's intervention in the matter. Consider these excerpts from the decision of the court written by Earl Warren, chief justice of the United States:

> Today education is perhaps the most important function of the state and local governments . . . it is required in the performance of our most basic public responsibilities, even service in the armed forces. It is the very foundation of good citizenship . . . it is doubtful that any child may reasonably be expected to succeed in life if he is denied the opportunity of an education. Such an opportunity, where the state has undertaken to provide it, *is a right* [emphasis added] which must be made available to all on equal terms. (*Brown v. Board of Education*, 1954)

In declaring education a fundamental right, like free speech or trial by jury, the court opened the door for it to consider the case as a constitutional matter and apply "strict scrutiny," which justifies the law being challenged within this context. As applied to school finance, the Fourteenth Amendment is used to require that the government treat all fairly. After *Brown,* funding equity questions had the potential to be raised to a higher level of consideration: a constitutional matter. In school finance cases it is routinely argued that the state system of funding treats different kinds of students differently, with no justifiable reason, i.e., rational basis. Thus, equal protection is denied in such systems.

But the courts are and have been reluctant to wade into the arcane world of school finance. They recognize the appropriateness of each state creating its own system of schools and the means to fund them. And, to the extent that these finance systems deliver a relatively equal education to all students, the courts will not get involved. The burden of proof rests with the complainants, who must show that the state is maintaining a system that treats different classifications of students differently.

Battles for Equity

Serrano v. Priest, a 1971 California case, was an early success using the equal protection clause argument. In this case the California constitution was also at the center. The California court determined that the school finance system treated a classification of students differently from other children. It was children in low-property-wealth school districts who were systematically shortchanged. This met the threshold for the court to determine there was a "suspect classification" involved and then apply the "strict scrutiny" standard. Now the California court had a compelling interest to review the matter in detail and was justified in its call for remedial action.

The school finance system in California relied too heavily on local property tax as a primary revenue source. The plaintiffs were able to prove that the California system of school finance favored children in property-rich school districts and limited the educational opportunities of children in property-poor school districts. The disparities were so great that even when some low-wealth school districts taxed themselves to the maximum allowed by law, they could not raise enough money to offer an education program that approximated the average California school district's spending.

It was demonstrated that the state funding system was not "fiscally neutral," that is, biased toward some school districts and students. Thus, the plaintiffs triumphed in their assertion that these children comprised an eligible group (suspect classification) and were entitled to protection from this unfair government treatment.

Although *Serrano* was a state court decision, thus having no precedent outside of California, the case sent shock waves across the country as state legislatures, governors and departments of education began to consider their funding scheme against *Serrano* legal standards. Many states saw advocacy groups gear up to assert like legal claims in their state. Equity in school funding vaulted to the top of the policy agenda across America.

As previously mentioned, the legislative and executive branches of government jealously guard their policy making prerogatives and are loath to have a court take over this role. Given the extent of prescriptive court interventions with regard to school desegregation cases during that era, it was natural to assume the courts would follow similar patterns in school finance cases. With this in mind school finance systems were modified or totally revamped by legislatures to achieve a closer match with the new conception of school finance equity.

The success of *Serrano*-type cases in other states and the policy adjustments to funding programs inspired in the wake of *Serrano* moved plaintiffs in Texas to attempt to have their school finance case heard in federal court (*San Antonio Independent School District v. Rodriguez*). Their argument was the same: children in low-property-wealth school districts had less money spent on their education than children in high-property-wealth school districts. The Texas system was not fiscally neutral. The relief sought was a more equitable distribution of funding among school districts in Texas.

In 1973 the U.S. Supreme Court, in a 5–4 decision, ruled in *Rodriguez* that the Texas funding program did not discriminate, and that the plaintiffs failed to demonstrate how children in low-property-wealth school districts made up a suspect class subject to discrimination. Furthermore, the court decided that education was not a fundamental right—therefore, strict scrutiny did not apply—and that no group was being discriminated against. It was also determined that there was a "rational basis" for the existing Texas school funding scheme. Thus, this effort to apply the legal perspectives of *Brown* and *Serrano* to school finance litigation was unsuccessful at the U.S. Supreme Court level.

In an ironic twist of history justice Thurgood Marshall wrote the dissenting opinion in the *Rodriguez* case. Earlier in his career and before his appointment to the U.S. Supreme Court, Justice Marshall was the lead attorney arguing on behalf of the plaintiffs in *Brown*. In an eloquent and detailed sixty-page opinion, he used many of the legal concepts and the court's actual words from the *Brown* decision in his argument. Here are some excerpts:

> The majority's holding can only be seen as a retreat from our historic commitment to equality of educational opportunity and an unsupportable acquiescence in a system which deprives children in their earliest years of the chance to reach their full potential as citizens. . . .
>
> In fact, if financing variations are so insignificant to educational quality, it is difficult to understand why a number of our country's wealthiest school districts, which have no legal obligation to argue in support of the constitutionality of the Texas legislation, have nevertheless zealously pursued its cause before this court. . . .
>
> In this case we have been presented with an instance of such discrimination, in a particularly invidious form, against an individual interest of large constitutional and practical importance. (*San Antonio Independent School District v. Rodriguez*, 1973)

The disparities outlined in the case showed how extremes in property wealth rendered revenue distributions of equal consequence. For example, property tax in one district could raise $585 per pupil on thirty-one mills, while another could only raise $60 per pupil on seventy mills. Furthermore, it was demonstrated that the state foundation program did little to mitigate funding differences. In fact, much of the state funding was spread on a flat grant basis. The defendants sought cover in the argument that the poor district got an adequate amount of money to run their schools, and that money doesn't matter, anyway, in educational outcomes.

In the aftermath of *Rodriguez*, plaintiffs would have to rely on state courts to plead their cases for school finance equity. And a couple of state courts, in Arizona and Michigan, even reversed themselves on previous *Serrano*-type decisions. But success was achieved in many other states through litigation using *Serrano* arguments. Furthermore, the legal arguments and rationales of the plaintiffs in *Rodriguez* were strong, and upon reflection, many state legislatures took it upon themselves to correct injustices in their state funding systems. Saleh (2011) presents compel-

---

Text Box 4.3 Three decades of significant school finance cases.

Serrano v. Priest, 5 Cal. 3d 584, 96 Cal. Rptr. 601, 487 P.2d 1241. (1971).
San Antonio Independent School District v. Rodriguez, 411 U.S. 1. (1973).
Abbott v. Burke, 495 A.2d 376, 390. (New Jersey. 1985), and subsequent decisions.
Rose v. Council for Better Education, Inc., 790 S.W.2d 186 (Kentucky. 1989).
Campaign for Fiscal Equity v. State of New York, 100 N.Y.2d 893. (2003).

---

ling arguments that evolved thinking about school finance today may indicate that the time is right for another *Rodriguez*-type case at the U.S. Supreme Court level.

Battles for Adequacy

Logic would dictate that battles for equity should be preceded by struggles for adequate funding for schools. And, in fact, this was the case. The fight for sufficient resources for schools emerged concurrently with the development of the school systems themselves. The historical record contains example after example of struggles with inadequately funded schools from the colonial era through the early years of the common school movement.

Founding fathers like Thomas Jefferson, Benjamin Rush and Daniel Webster pleaded for tax-supported schools. During the nineteenth century, Horace Mann, Henry Barnard and many other common school advocates exhorted state school boards and legislatures to provide adequate funding for decent schools. In those early days, adequacy was concerned with dilapidated facilities, basic equipment, salaries, short school terms and the general capacity of the system to meet its purpose.

Adequacy battles in the modern era often take on the same concerns. Although much more sophisticated and elaborate, legal arguments in recent decades over adequacy in funding still look at such things as facilities, teacher salaries, curriculum, materials and equipment. But results, or student outcomes, play an important role in these litigations as well. The assertion that education is a foundational institution essential to the success of society and the individual is still put forth. Issues of fairness and constitutionality are also argued. However, since the accountability movement, ushered in by the 1983 seminal report *A Nation at Risk,* the results of the education system are of central concern. During the past three decades, more than half the states have been challenged by adequacy lawsuits, and three-quarters of the time the plaintiffs have prevailed.

The blockbuster case for the modern era of adequacy litigation comes from Kentucky, *Rose v. Council for Better Education, Inc.* The 1989 state supreme court decision was broad in scope and precise in detail. The court, in reviewing the adequacy question, not only looked at finances across the state and between school districts but considered the entire Kentucky pre-collegiate education system. Reflecting on the language of the Kentucky state constitution, the court declared the education system was not "efficient" and then proceeded to outline what kind of system would be efficient. It also determined that the system was not "adequate" and also specified what adequacy meant. The court enumerated its meaning in very precise terms. Consider these excerpts from the decision:

> A child's right to an education is fundamental under our Constitution . . . The common schools are free to all . . . substantially uniform . . . provide equal educational opportunities to all Kentucky children, regardless of place of residence or economic circumstance . . . its goal is to provide each and every child with at least the seven following capabilities:
>
> i. Sufficient oral and written communication skills to enable students to function in a complex and rapidly changing civilization;
> ii. Sufficient knowledge of economic, social, and political systems to enable the student to make informed choices;
> iii. Sufficient understanding of government processes to enable the student to understand the issues that affect his or her community, state, and the nation;
> iv. Sufficient self-knowledge and knowledge of his or her mental and physical well being;
> v. Sufficient grounding in the arts to enable each student to appreciate his or her cultural and historical heritage;
> vi. Sufficient training or preparation for advanced training in either academic or vocational fields so as to enable each child to choose and pursue life work intelligently; and,
> vii. Sufficient levels of academic or vocational skills to enable public school students to compete favorably with their counterparts in surrounding states, in academics or in the job market. (*Rose v. Council for Better Education, Inc.*, 1989)

This case ushered in a total redesign of the education system in Kentucky. The fact that the court specified outcomes required that a system of monitoring educational results also be established. Thus, Kentucky was one of the first states to build a statewide student assessment system, which anticipated the federal No Child Left Behind Act by well over a decade. Uniformity requirements caused substantive revision and expansions of curriculum. Everything from school governance, to facilities, to teacher salaries came under scrutiny.

The financial impact of the case was also enormous for Kentucky. Resources for schools increased by 30 percent within a few years of the decision, tax revenue for schools grew by over $1 billion annually and disparities in funding between districts decreased by 37 percent. The Kentucky Education Reform Act (KERA), which was the vehicle for the education redesign, undertook a comprehensive sweep of the education system (Hess, 2006).

Contemporary Adequacy Cases

We turn now to the *Abbott* cases in New Jersey (*Abbott v. Burke*, 1985), yet another sweeping but more convoluted group of adequacy cases. In a series of no fewer than ten decisions of the court ranging over twenty years, New Jersey has struggled to resolve issues of fairness and sufficiency in its education and school finance system. The *Abbott* legal battles underscore the complexity and political dynamics associated with high stakes school finance reform. Despite the addition of billions of dollars to the funding scheme and extensive changes to the education system, the battle grinds on.

In *Abbott*, one sees the conflict between the courts and the legislature, the legislature and the governor, and interest groups of all stripes defending, challenging and advocating for their vision of adequacy and equity. The age-old battle for resources and values unfolds in the courts, the legislature, the governor's office and the media. Alexandra Greif (2004) in her paper *Politics, Practicalities, and Priorities: New Jersey's Experience Implementing the Abbott V Mandate* traces the intricate legal and political moves made by the various stakeholders involved in this expansive case.

The *Abbott* cases encompassed everything from preschools, to school facilities, to tax reform. Governors and legislators came and went, and the struggle continues. The *Abbott* cases led to the identification of a group of school districts targeted for special attention. These districts gained unique consideration programmatically, financially and administratively. Other litigators, school adequacy advocates, tax limitation groups, scholars, professional organizations and a host of others have dissected and deconstructed the legal arguments and remedies that have come down from the *Abbott* courts. No final resolution has been achieved.

Not to be outdone by its next-door neighbor, New York State has grappled with its own adequacy court fights for over a decade. Originally brought forth by plaintiffs from the New York City schools, the *Campaign for Fiscal Equity v. State of New York* was eventually joined by other school districts who perceived themselves to be underfunded by the New York state school finance approach. Although the plaintiffs chose to use the name "equity" for their group, the case has been about adequacy of funding. Here again, the courts ruled in favor of the schools, this time to the tune of an additional 43 percent or $5.6 billion per year for New York City alone (Rebell, 2008). But as has been the case in many other adequacy cases, court rulings and action by the legislative and executive branches do not always function in sync.

Years have passed and additional court rulings have followed, but it was changing politics that added impetus to the court mandates. By 2006 changes in the political makeup of the state capital, Albany, saw substantial amounts of new funding headed to schools. As is often the case, all schools across the state saw increases in financial resources. Major initiatives for new facilities and rehabilitation of existing facilities have come to life in New York City. Expanded curricula and educational programs are also a result of the increased appropriations.

There have been many significant adequacy cases in other states, for example, *Edgewood Independent School District v. Meno* in Texas and *Coalition for Adequacy and Fairness in School Funding v. Chiles* in Florida. Each has contributed to the canon of legal thinking about adequacy of school funding. New cases emerge each year and old cases are revisited. However, the current condition of the economy across the globe and in the individual states has precipitated a crisis in funding for education.

Cutbacks in funding are the norm, even in states with court mandates to address adequacy concerns.

Summary

The battles for school funding are a part of the education landscape and have been since the beginnings of the common schools. Constitutional questions, federal and state, are the basis for much of the education finance litigation. Fairness and sufficiency have been the focus of these legal fights. The courts continue to be a central player regarding questions of equity and adequacy. The political dynamics among the legislative, executive and judicial branches of government ultimately determine the course of legal mandates regarding equity and adequacy. The educational results of these lawsuits, win or lose, remain controversial (Hanushek and Lindseth, 2009; Rebell and Baker, 2009).

Equity cases contributed much to equalizing funding among school districts and expanding the legal thinking about standards of fairness with regard to the distribution of resources and the burden of raising funds from taxpayers. Unlike the equity cases, which generally restricted themselves to funding issues, adequacy cases reach into the substance of the education system itself. Everything from curricular offerings to teacher quality is touched. Often the plaintiffs prevail in this kind of litigation, but even when they don't, states frequently modify resource allocations for schools.

Chapter 5 provides a more in-depth view of equity and adequacy in education finance. Various measurement techniques used by scholars, economists and social scientists are explained. Multiple perspectives on the issues are presented.

References

Abbott v. Burke, 495 A.2d 376, 390. (New Jersey, 1985).

Alexander, K., and Alexander, D. M. (2009). *American public school law* (7th ed.). Belmont, CA: Wadsworth, Cengage Learning.

Brown v. Board of Education, 347 U.S. 483, 74 S. Ct 686. (1954).

Campaign for Fiscal Equity v. State of New York, 100 N.Y.2d 893. (2003).

Coalition for Adequacy and Fairness in School Funding v. Chiles, 680 So.2d 400. (Florida, 1996).

Committee for Educational Rights v. Edgar, 641 N.E.2d 602 (1994), 267 Ill. App. 3d 18, 204 Ill.

DeRolph v. State, 78 Ohio St.3d 193, 677 N.E.2d 733. (1997).

Eastman, J. C. (2007). Reinterpreting the education clauses in state constitutions. In West, R., and Peterson, P. E. (Eds.), *School money trials: The legal pursuit of educational adequacy* (pp. 55–74). Washington, DC: The Brookings Institution.

Edgewood Independent School District v. Meno, 917 S.W.2d 717, 732. (Texas, 1995).

Greif, A. (2004). Politics, practicalities, and priorities: New Jersey's experience implementing the Abbott V mandate. *Yale Law and Policy Review, 22*(2), 623.

Hanushek, E., and Lindseth, A. (2009). The effectiveness of court-ordered funding of schools. *Education Outlook, 60*. American Enterprise Institute for Public Policy.

Hess, F. M. (2006). Adequacy judgments and school reform. In West, R., and Peterson, P. E. (Eds.), *School money trials: The legal pursuit of educational adequacy* (pp. 159–94). Washington, DC: The Brookings Institution.

Lujan v. Colorado State Board of Education, 649 P.2d 1005. (Colorado, 1982).

Marbury v. Madison, 5 U.S. (1 Cranch) 137 (1803).

National Access Network (2007). *Litigations Challenging Constitutionality of K–12 Funding in the 50 States.* Retrieved from http://www.schoolfunding.info/litigation/In-Process%20Litigations.pdf

National Commission on Excellence in Education (1983). *A nation at risk.* Washington, DC: U.S. Department of Education.

Plessy v. Ferguson, 163 U.S. 537, 16 S. Ct. 1138. (1896).

Rebell, M. A. (2008). Equal opportunity and the courts. *Phi Delta Kappan, 89*(6), 432–39.

Rebell, M. A., and Baker, B. D. (July 8, 2009). Assessing "success" in school finance litigations. *Education Week, 28*(36). Retrieved from http://www.edweek.org/ew/articles/2009/07/08/36rebell.h28.html

Roberts v. The City of Boston, 59 Mass. (5 Cush.). (1850). Retrieved from http://brownvboard.org/research/handbook/sources/roberts/roberts-210.htm.

Rose v. Council for Better Education, Inc., 790 S.W.2d 186. (Kentucky, 1989).

Saleh, M. C. (2011). Modernizing *San Antonio Independent School District v. Rodriguez*: How evolving Supreme Court jurisprudence changes the face of education finance litigation. *Journal of Education Finance, 37*(2), 99–129.

San Antonio Independent School District v. Rodriguez, 411 U.S. 1. (1973).

Serrano v. Priest, 5 Cal. 3d 584, 96 Cal. Rptr. 601, 487 P.2d 1241. (1971).

Stuart v. School District No. 1 of Village of Kalamazoo, 30 Mich. 69. (1874).

Equity and Adequacy in Education Finance   5

Aim of the Chapter

IN THIS CHAPTER THE DEFINITIONS, CONCEPTS AND issues associated with equity and adequacy in school finance are presented. The various dimensions and nuances of the topics are outlined and the theoretical and practical aspects of the issues are discussed. Several analytic methodologies and tools are presented to assist in formulating personal perspectives about these topics.

Introduction

Since the beginning of the public schools, questions about the amount and distribution of resources to fund education have been at the center of discussion, controversy and debate. To this day the issues of equity and adequacy emerge in conversations from the nation's capitol to the neighborhood schoolhouse. Parents, teachers, administrators, school board members, legislators, governors, business people and advocacy groups are all concerned with questions of equity and adequacy for school funding. These are perennial issues that have evolved into ever higher levels of esoteric discourse and legal battles. Yet, despite the movement toward more technical and sophisticated analyses of these issues, there is no shortage of opinions from all quarters about the equity and adequacy of funding for schools.

Equity

Like beauty, equity is often in the eye of the beholder. This is because one's perspective as a student, a teacher, a parent, a legislator, a taxpayer or a school administrator colors one's opinion about equity. For the average citizen, equity in school funding means equal funding for all schools. But this concept of equity started to lose favor among scholars, policy makers and school leaders in the early part of the twentieth century as thinking in school finance theory gained refinement. Early studies showed that equal funding—through mechanisms like equal tax levies, lump sum grants, flat grants to school districts, or equal per pupil funding—did not result in equal programs and services to students and teachers (Cubberley, 1906; Updegraff, 1922; Strayer and Haig, 1923; Mort, 1924; Johns, Morphet and Alexander, 1983).

The intersection of state goals for education—in the form of desired outcomes for students, the model of education service delivery envisioned by the state (i.e., a graded one through twelve system) and the resources provided to individual school districts—repeatedly emerges as a point of contention throughout the historical record. Great emphasis was placed on standardizing the education system within states during the late 1800s and early 1900s. However, equal funding, it was determined, led to unequal programs, services and results.

A big reason why equal funding leads to unequal school systems is that individual school districts are not all alike. They do not have the same number and types of students with the exact same educational needs. They do not have the same staff with identical levels of experience, training and expertise. School districts are not established in identical sizes and geographic settings. They do not all have the same local economy and resources upon which to draw. Numerous other reasons contribute to the variation among school districts. Much of this variation is highly predictable, while in other cases it is unique to an individual school district.

Policy makers have struggled with questions of equity in school funding throughout the history of public education. Until the latter half of the twentieth century, in most of the states, the majority of school money came from local sources, generally property tax. As is the case today, school districts operated within a state system of education. The state would declare the desired outcomes and the means of education, e.g., eight years of schooling for children between the ages of six and sixteen; in the subjects of reading, writing, arithmetic, history and civics; schools to run for a minimum of ninety days; teachers must be certified by the state; local property to be taxed at a particular rate for the support of schools. School facilities tended to be matters of local concern with regard to financing, although the state often did not hesitate to set standards for school buildings.

Because the system was built on mostly local dollars, local wealth was the paramount determinant of the resources available to the school district, and local wealth was also highly correlated to school results. For example, poor districts might struggle to meet minimum state requirements, whereas wealthier communities could provide vastly more expansive programs. The parochial perspective, fostered by most school funding coming from local sources, reinforced the practice of school districts looking internally for their standards and benchmarks. This also contributed to disparities among school districts.

Several factors contributed to the longevity of the practice of inequality posited as equality. Through the nineteenth and early twentieth centuries, school districts were overwhelmingly small and rural. This pattern reflected the distribution of the population and the realities of travel. Local rural travel was often more difficult than

Text Box 5.1 A Very Wealthy School District

Sometimes the variations that exist among school districts are so great that the differences border on the ridiculous. As an example, consider the case in one Midwestern state with a long history of wide disparities in school funding. One of the more extreme examples arose as a result of the state's overreliance on property tax to fund its schools. Depending on the school district's property wealth, calculated as assessed property value per student, one district could have as much as five times the resources as the least wealthy school districts.

A policy that further aggravated the inequitable situation required school districts to use local property tax as the means of financing new construction for schools. In other words, districts would have to ask the local voters to approve higher taxes in order for the school district to issue bonds (debt) to raise money to pay for new facilities. The state had a meagerly funded grant program to help poor school districts, but this program had a long waiting list and rarely provided enough assistance to help the poorest school district supplement local resources to the point of affording new construction.

In this state there was one small-town school district that had so much assessed value within its boundaries that it was able to build a new state-of-the-art high school, fully equipped, without issuing one dollar's worth of debt. The school district was collecting so much money through its operating budget tax levy (the amount authorized by state statute) that it was able to save cash over several years and pay for the new high school with cash. Most school districts across the country have to sell bonds, which are typically paid back over a twenty-year period, to finance the construction of new schools.

But the story does not end there. About 20 percent of the state's school districts had such low assessed valuation that they could not afford to maintain their existing facilities, much less build new schools. The vagaries of agricultural commodity prices and the decline of manufacturing in these school districts contributed to these pockets of economic depression. As one traveled around the state, the apparent contrasts between property-rich and property-poor school districts in this state were stark.

inter-city travel, due to poor roads and limited means of transportation. Thus, the icon of the one-room schoolhouse, to which students would trod everyday, dominated the landscape. School districts proliferated as each neighborhood in town and clusters of families throughout the countryside formed one.

With the Industrial Revolution in America, the urban centers expanded greatly and larger school districts emerged from smaller ward-based ones in the cities. Discrepancies between poor rural schools and well-funded small-town and large-city schools widened. Along with the shifting population came reapportionment in the state legislature. As the political power shifted from rural and agricultural centers to

Picture 5.1   The one-room school is an icon of American education.

urban and industrial centers, so would the resources for schools. This was a pattern repeated in many states across the country. By the late twentieth century the power had shifted again, this time to the suburbs.

To the credit of a few states, these discrepancies were addressed through the political process and the leadership of civic authorities, chief state school officers, legislators and governors. But overall, nineteenth and early twentieth century politics are not known for magnanimous decision making or fairness. Discrepancies among the schools in many states were stark. The pattern was fairly common: communities with low property wealth tended to have inferior schools to a much greater extent than schools in communities with high property wealth. Furthermore, rigid segregation patterns in housing during this era served to further exacerbate this situation for poor whites and minorities.

For the most part the discrepancy derived from inadequate resources, because of the school district's inability to collect sufficient money, even with authority for the same level of taxation as the wealthier districts. In other words, even when rich and poor districts were allowed to tax at the same rate—say, for example, by a state-established tax rate—the amount of money they could each raise would be vastly different. In some instances communities would be allowed to exceed the state minimum-required tax levy, but this "solution" would often serve to make matters worse, as rich school districts had an easier effort to raise more dollars.

Evolution of Equity

The concept of equity in school finance has developed over time to a more nuanced and sophisticated formulation. As mentioned above, the historical record tells us that equity tended to be viewed as equal by policy makers as they distributed lump sum and flat grants to communities to encourage the establishment of schools. Criticisms of this approach were immediate and ongoing. Proportional distribution was seen as a major advance, and the idea of per student funding and per student revenue gained in popularity (Ramirez, 2003).

Some states and school districts, for example, in an effort to tie costs to revenue generation, introduced the concept of "teacher units" to the equation. Under this methodology the number of teachers required to serve the student population was determined and this figure was used as the basis for determining revenue needs. But this practice also proved inadequate, since students don't show up at the schoolhouse in perfectly distributed age groups and grade levels. It also assumes students live in convenient and equal clusters near schools.

A look at state legislation around the country during this era will reveal an array of procedures aimed at mitigating funding disparities among school districts. However, as is the case to this day, the political interests of various groups—i.e., suburban, urban, rural, industry, labor, agriculture, white, minority, social class— can also be seen in these statutes. Among the many definitions of politics, one in particular seems appropriate to this discussion of equity; to paraphrase "Boss"

Tweed of Tammany Hall in New York City—politics is the business of deciding who gets what."

Multiple Concepts of Equity

Today equity is viewed as a multidimensional concept where numerous factors interact within a dynamic tension. Equity has historically been judged between school districts, but old and simple formulations of equity can no longer be sustained. Progress in theoretical thinking about school finance and legal pronouncements from the courts have established a broader vision of equity. Below are some of the more contemporary perspectives on equity:

STUDENT FOCUS   This view takes the position that each student, regardless of his or her school district's location or the wealth of his or her community, should have an equal educational opportunity. In other words, every student should have access to the same educational programs, number, quality, and types of teachers, instructional resources and facilities. Great efforts were made across the country in the late nineteenth and early twentieth century to standardize the education service delivery system. A major impetus for this effort was to achieve programmatic equivalence among the school districts in individual states. Today it is recognized that differences among school districts require different financial resources to achieve desired educational standards.

TEACHER FOCUS   This perspective asserts that if resources are distributed equally among the school districts, the real test of equity is whether the teachers have the resources available to them to educate the children in their charge. This theory of equity proffers the idea that different geographic settings and student populations will cause teachers to need different resource levels to educate the children in their schools. Thus, the measure of equity is whether the teachers have what they need to get the job done, e.g., prepare students for entry to the state university, the labor force, the military, or post-secondary vocational training.

TAXPAYER FOCUS   This facet of equity takes the position of the taxpayer in the school district. It seeks to set a fair level of tax effort among school districts. For example, property-wealth disparities among school districts can result in unequal tax burdens among school districts. Consider the scenario in which two school districts are compared and residential property tax is the basis of school funding:

> District A has an average assessed valuation on residential property of $100,000, while neighboring school district B has an average assessed value on residential property of $50,000. Both districts have the same size student population and number of taxpayers. The state requires that all school districts raise at least one-half of the per pupil operating funding through a local tax levy. In this example, taxpayers in district B will have to pay twice as much as taxpayers in district A to raise the same amount of per pupil funding.

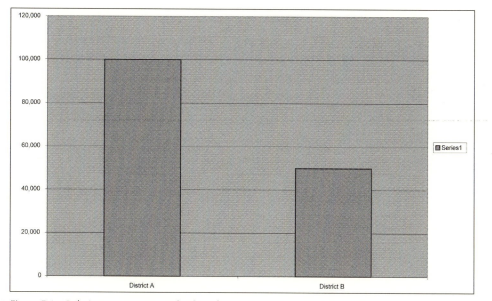

Figure 5.1   Relative average assessed value of property in two school districts.

ASSESSMENT FOCUS   How property is assessed can also affect equity. When assessments of real property vary widely from market values, i.e., actual selling price, unfairness spreads through the system. In states where local tax assessors are more politicians than professionals, problems can occur. It is also a problem when taxpayers have few options to appeal assessments on their property. Here is an illustration: a house sells for $300,000 in school district A and is assessed for tax purposes at $30,000. A similar house in school district B also sells for $300,000 and is assessed for tax purposes at $60,000. Even if the state requires a uniform tax levy for schools, the district B homeowner will pay 50 percent higher taxes. Thus, achieving a fair and uniform assessment system helps to promote equity for taxpayers.

WEALTH FOCUS   This orientation seeks to consider the ability of the community to support its schools beyond considering only property values. Not all states tax property in a uniform manner. Often, residential, commercial, agricultural, industrial, forestry, mining and utilities are taxed at differing rates and the taxes are designated for the support of different public purposes. Just looking at the value of residential property within a school district as a measure of wealth may mask a truer picture. Equity requires a broader conception of wealth within the community. For example, income is also an important consideration, since the relationship between income and property value can vary greatly. Think of the elderly widow on a fixed income who lives in a house that is worth ten times what she paid for it forty years ago, and compare her situation to her neighbors, two young professionals with no children. Income and demographics can be better indicators of wealth within a school district.

Text Box 5.2 Avoiding taxes.

When I was a little boy my family lived on a farm. It was a wonderful part of the state and I had lots of aunts and uncles who lived on farms as well in the surrounding area. I remember how every spring my dad and his brothers and sisters would move a lot of the farm equipment and implements from one farm to the next and put them up in the barns. They wouldn't use the equipment, just store it in the barns. It looked to me like some kind of musical chairs with tractors. When I was a bit older I got to help out and drive some of the equipment. It was kind of festive and fun. It seemed like a lot of the farmers in the area did the same thing.

It wasn't until I was a young man and had returned to my hometown that I finally learned what the "tractor musical chairs" was about. While lunching with my dad at the local Elks club Mr. Jones, the county assessor, stopped by our table to say hello. During the brief conversation with my dad, Mr. Jones mentioned that he would be in our neighborhood next week assessing personal property on the farms in the area. After he left my dad turned to me and asked, "Are you going to be here next week so you can help us move tractors to your Uncle Bill's place?"

COMMUNITY FOCUS   This looks more toward the outcomes of the system. Here the notion is that each community should have schools that provide an equivalent level of education to their graduates. This analysis considers, for example, whether employers have a properly educated workforce, whether graduates are prepared for the state universities or whether the school district's reputation adds or detracts from property values.

There are many ways to view equity, and each focus requires a set of unique adjustments in order to achieve the desired result. The concept of equity has developed over time and continues to be refined. In the section below, some of the methods of determining equity are explained.

Measuring Equity

As with any argument, the debaters appeal to rational and objective criteria to support their cause. The same is true in making the case for equity in school funding. Much of these analyses are rooted in statistical procedures. At a minimum these objective criteria serve as a starting point to understand the nature and extent of funding differences between school districts. But keep in mind that while the measures applied to the debate will show differences, differences do not automatically prove inequity.

MULTIPLE PERSPECTIVES OF EQUITY   School finance draws on scholarship from economics, law, political science, sociology and other fields in an effort to find tools that will help with complex problems like equity in school funding. One of the more important works was authored by Robert Berne and Leanna Stiefel in 1984, in which they introduced several key concepts about funding equity. The idea of horizontal equity, vertical equity and fiscal neutrality are examples. Berne and Stiefel recognized that merely measuring across school districts to judge the equality of revenue distribution—horizontal equity—was inadequate. They understood that the challenge in school finance was to find ways to measure equity among very different school districts, vertical equity. Thus the "equal treatment of equals," or *horizontal equity*, and the "unequal treatment of unequals," or *vertical equity*, were coined as expressions of a broad view of equity. The concept of *fiscal neutrality*, where local wealth is not an advantage in the distribution of funding, was another important idea as a measure of equity.

Here are several more ways equity in school funding is measured:

DISTRIBUTION VIEW   This approach seeks to display all school districts in a state along a continuum (the bell curve) from lowest funded to highest funded. School districts that fall beyond one or more standard deviations from the mean are subject to scrutiny. Sometimes statisticians will use other kinds of standard scores to facilitate comparisons, e.g., quintiles, deciles or stanines. An important purpose of these methods is to identify outliers for investigation. In using this method, it is also common to restrict the range of the sample under study by dropping the extreme ends of the distribution. This helps to gain a truer picture of the state while eliminating the anomalous school districts, found in every state, that will always need special treatment.

A number of statistical formulas and adjustments, such as the *federal range ratio*, are used by school finance experts and economists to account for the extreme variation among school district, e.g., size, wealth, cost, special populations and other statistical extremes, which contribute to problems of scale when making comparisons within a range of school districts. Consider, for example, comparing school districts in New York state and having a New York City school district of one million students; California has Los Angeles, Illinois has Chicago, and so forth. Conversely, many states will also have very small school districts; Colorado, for example, has a school district with fewer than 100 students.

Extremely large- and small-size school districts within a state will distort the range, mean, distribution and so forth. Anomalies other than size will invariably be found among school districts in a state. However, understanding the range or distribution of funding across school districts in a state remains an important measure. The National Center for Education Statistics (NCES) website[1] offers an example of a state-by-state comparison of current expenditures per pupil at the 5th percentile, median and 95th percentile cutpoints and the federal range ratio.

RELATIONAL VIEW   This view strives to show how differing aspects of funding and school district circumstances affect some established formulations of equity. Thus, school finance scholars, economists and policy leaders will use any number of statistical calculation methods to display relationships among variables that affect equity.

Correlation will show how one variable reacts in relation to another, e.g., local school district wealth and expenditures per pupil. *Regression analysis*, on the other hand, will exhibit the direction and power of multiple variables, e.g., wealth, expenditures per pupil, district size and percent of special population. The overall effect of the interaction of multiple variables and the influence of an individual variable on the whole can be determined with this methodology. This method can help explain funding differences among school districts.

Various other statistical methods to assess equity using formulas that seek to account for relationships among factors that contribute to or detract from equity are used. Many of these processes are derived from economics, finance, science and other technical fields. Here is a list of the more commonly used approaches: the *Lorenz curve* graphically displays the relative inequality between two variables; *coefficient of variation* shows the degree of variation from the mean among different data sets, e.g., rural and urban school districts; the *Gini coefficient* quantifies the degree of dispersion from the mean as a way to determine inequality; the *McLoone Index* is concerned with those districts that fall below the mean; the *Theil Index* can measure inequality within and among subgroups of school districts or schools (Downes and Stiefel, 2008). Hussar and Sonnenberg (2000) offer a more full explanation of the above measures on another page of the NCES website.[2]

Even more sophisticated measures of relationships are used to gauge equity. *Hierarchical linear modeling* allows for analysis that judges the effects of equity vertically through multiple layers of the system, e.g., school, district, state (Bryk and Raudenbush, 1992). While often applied to studies of efficiency in spending, *data envelopment analysis* (Cooper, Seiford and Zhu, 2011) makes it possible to bundle several variables in an effort to estimate their effect on a result. Similarly, the *stochastic frontier analysis,* as the name implies, bundles variables in order to compare how individual cases match up against the calculated "frontier" for a group of cases, i.e., school districts.

Concepts of Funding Adequacy

How much? That is the fundamental question of adequacy in education funding. It is a question that has been asked throughout the history of public education, and it is a question that is asked by parents, taxpayers, school board members, superintendents, legislators and governors every year in every state. Somewhere between having no money and having an unlimited supply of money is a theoretical figure that meets adequacy. The term "theoretical" is used because there is no absolute measure of adequacy across the nation, within states or even within school districts.

Each state has established its education system on a framework of constitutional provisions, state statutes, court decisions and policies that sets parameters for expected results. But these parameters can range from basic skills for survival in modern society to fulfilling individual human potential. Furthermore, the schol-arly, legal, educational and social frontiers are constantly evolving and with them the parameters of expectations for schools. Over the past several decades numerous state courts and state legislatures have wrestled with questions of adequacy, and ultimately they are the decision makers in these matters.

In the modern era, adequacy has been at the center of much litigation and much debate among policy makers. This attention to adequate funding was stimulated in recent decades by policy decisions to explicitly define academic standards related to curricular content and student academic performance. The content and perfor-mance standards were further linked to a system of student assessment. Thus, ex-pectations were clearly articulated and measured. Within this theoretical construct, adequate funding serves to help schools help students meet the standards.

Standards, assessment and adequate funding can be compared to a three-legged stool. Clear expectations, measured results and sufficient funding each support a system that gets all students to high levels of learning. Jennifer O'Day and Mar-shall Smith (1993) articulated a theory of improved learning in their seminal essay "Systemic School Reform and Educational Opportunity." State and national policy makers applauded the accountability measures of the theory for the most part but balked at the vast unknown of determining adequate funding. To many policy makers, the idea of enshrining adequacy in legislation along with standards and as-sessment proved to be a leap of faith too wide to take.

Defining Adequacy

Policy analysts, lawyers and school finance scholars have been engaged in the search for better ways to determine adequacy. Unlike equity, which has a century-long repository of scholarship bolstered by sophisticated statistical calculations, adequacy studies tend to be of a more recent vintage. Downes and Stiefel (2008) present four methods of measuring adequacy:

PROFESSIONAL JUDGMENT   This method typically uses a panel of experienced and respected education professionals and experts to design a model school, which will serve a particular mix of students, e.g., regular, special education, English lan-guage learners, gifted and so forth. The theoretical school is designed with all the programs and support services necessary to help all students succeed. The model is based on the best judgment of the experts without using extravagant additions. Often, the process calls for the model school to be vetted among other experts, educators, policy makers, community members, etc. The cost of the final model is calculated based on current trends in expenditures. These cost figures are then extrapolated to the broader system to determine adequate funding for all school districts across the state.

SUCCESSFUL DISTRICT   This procedure seeks to identify benchmark school districts—i.e., districts that are achieving academic success, but within a spending range that does not deviate too far from the median. These exemplar districts are then analyzed to determine the organizational staffing, services and programs they use to achieve their academic results. The cost model built from this analysis is used to establish a per pupil figure. This per pupil figure serves as the basis of funding, subject to adjustments related to the unique circumstances of each school district such as size, demographics or historical spending patterns.

WHOLE SCHOOL   Based on research from school reform initiatives, successful whole school design models are applied to existing schools within a school district. The costs for adopting the whole school design model are determined and applied to all the school districts across the state. This methodology adds funding based on the cost of bringing existing schools up to the ideal. Revised curriculum and professional development are often part of the added expense.

SCHOOL DISTRICT COST FUNCTION   In this technique a formula is used to discern the relationship between inputs and outputs. Inputs are costs for things like teachers, materials, supplies or programs. Outputs are typically measured by things like test scores, attendance rates and graduation rates (Hanushek, 1986). The cost function approach comes from economic theory and purports to render a cost function for each outcome and type of student: elementary, high school, special needs, etc. In other words, raising test scores for eighth-graders by "X" percent will cost "Y" dollars.

These methods for determining adequacy represent the current state of the art. Each model has its limitations. Personal bias is an obvious caution. The challenges of accounting for the nuances present in each school and school district are for-midable. Straight-line projections from calculated averages are always a concern as they reach school districts further away from the mean on any particular measure. The search for a template to apply across the state is elusive and might not even be wise. Yet, the question remains critically important: How much?

Other Concerns about Equity and Adequacy

The chapter thus far has been concerned with issues of equity across school districts and adequacy within a state system of education overall. But these same issues can come into play when looking at the amount and manner of resources allocated within school districts. Concerns about equity and adequacy within a school dis-trict are commonly expressed by those individuals closest to the student, namely teachers and parents.

Grade Levels

Unified school districts, i.e., PK–12 districts, developed later in the history of pub-lic education. High schools came on the scene slowly in the late nineteenth and early twentieth century, with junior high schools and middle schools coming even later. Some states still hang on to these historical artifacts and run separate elementary and separate high school districts. The more common mode is the combined or unified school district.

Part of the reason for the unequal distribution of resources among the various levels of schooling is the fact that decisions about funding within the district are, for the most part, the purview of the local board of education. A second reason is tied to efficiency.

Elementary schools tend to be more efficient than the other levels because they require less specialization. Middle schools have expanded curricula and more programs, so they have a greater mix of personnel. High schools are the least efficient of the three. High schools have a more complex mission, with multiple outcomes for students. As such, they provide a broader array of curricular and extracurricular offerings and programs. This adds to the demand for specialists of all types.

High schools also tend to be less efficient when they schedule smaller class sizes to accommodate the range of student interests and abilities. Unlike elementary schools, where common practice sees students of various levels of learning clustered by age, high schools routinely group students according to their readiness for the subject matter, i.e., algebra I and II, or English I through IV. Thus, per pupil expenditures tend to rise with grade levels. In a state where each student is funded at the same rate, claims of unfairness are sometimes made.

Transportation

School districts come in all shapes and sizes, and with varying geography. There are school districts in the West where a school board member will have to drive over 100 miles to get to a school board meeting. In other states, mountains, man-made structures such as airports or interstate highways, bodies of water or other topographical features will inhibit easy travel. One consequence of these variations in size and topography is that costs for student transportation will also vary greatly.

In states where transportation costs are recognized within the general operating budget, and without an equalizing factor, the results can be quite unfair. A compact school district with few geographic challenges will often have smaller transportation costs than a large-area school district with a similar size student population that is widely dispersed and has many topographic challenges. In these two cases the cost for the less compact school district for transporting students will be greater overall and the cost per individual student will be more.

Some states run separate transportation reimbursement formulas to mitigate this equity issue. In other states school districts are free to seek a *mill levy increase* (tax rate increase) with voter approval to supplement transportation costs. But, the latter arrangement only exacerbates inequity among rich and poor school districts.

Running buses loaded with forty children for five or six miles is more economical than servicing bus routes of ten to fifteen miles for ten or fifteen students each. Size and geographic factors are directly related to fuel, personnel and equipment costs. Two school districts with similar size student populations can have vastly different transportation costs. When transportation costs are not equalized, they adversely affect equalization of funding across school districts overall. Two districts that are funded at the same per pupil amount, under an otherwise equalized state funding formula for their general operating budgets, will quickly become unequal because of unequal transportation costs. For example, two districts are funded at $8,000 per pupil: district A averages $300 a year per student for transportations, while district B spends $900 per student. The net result is that on day one of the school year, each child in school district B will have $600 dollars less for his or her education that year than the children in school district A.

Facilities

In many states, the responsibility for raising money to acquire land and build schools is entirely the responsibility of the local school district. Other states offer minimal help through grants or matching funds, and some states control the funding and construction of new schools very closely. The clearest forms of inequity emerge when a school district must tax its residents to build new schools with no help from the state. School districts with greater assessed valuation generally have easier circumstances for raising the money for school construction than school districts with lower assessed valuation.

In such states it is common to see great disparities in facilities between communities. School districts with a large and growing tax base will have new and well-maintained schools. In contrast, school districts with a small and declining tax base will have older, dilapidated facilities. For the low-wealth districts the problem is exacerbated because old buildings tend to be more expensive and harder to maintain.

Interstate Equity

In more recent years, issues about spending for education, as compared between the states, has been raised as a cause for concern. The range in spending from the lowest- to the highest-spending states is vast. New Jersey, for example, spends more than twice as much as Utah, and Wyoming spends even more (Education Week, 2012).

Arguments about this issue divide into two camps. One group contends that since PK–12 education is a state function and not part of a national system, disparities between states are irrelevant. States are free to spend as much or as little as they choose. Members of the other camp assert that a more global view is needed. They see groups of students—for example, children from low-income homes—as being uniformly disadvantaged by inadequate funding, regardless of the state in which they live. Recently, comparisons of state spending within geographic regions of the nation have garnered some consideration in funding equity and adequacy litigation.

Summary

Discussions and debates about school funding take place at all levels of society. From the local school board to the national government in Washington, D.C., equity and adequacy are a focus of attention. The study of these topics has evolved over time and gained ever greater degrees of sophistication.

Determining equity and adequacy has become an esoteric undertaking involving statisticians, economists and school finance experts. Despite all the expert and technical analysis, questions about equity and adequacy are frequently arbitrated by the courts. Even these decisions are revisited as new legal perspectives and methods of calculation are brought to bear. Ultimately, however, the questions remain the same: What is fair, and how much is enough?

Chapter 6 investigates how states collect and distribute funding for schools. The chapter illustrates how the theories regarding equity and adequacy, discussed in this chapter, are addressed in state funding systems.

Notes

1. http://nces.ed.gov/pubs2009/revexpdist07/tables/table_04.asp
2. http://nces.ed.gov/pubs2000/2000020.pdf

References

Berne, R., and Stiefel, L. (1984). *The measurement of equity in school finance: Conceptual, methodological and empirical dimension.* Baltimore: Johns Hopkins University Press.

Bryk, A. S., and Raudenbush, S. W. (1992). *Hierarchical linear models.* Newbury Park, CA: Sage.

Cooper, W., Seiford, L., and Zhu, J. (2011). *Handbook on data envelopment analysis.* New York: Springer.

Cubberley, E. P. (1906). *School funds and their apportionment.* New York: Teachers College Press.

Downes, T. A., and Stiefel, L. (2007). Measuring equity and adequacy in school finance. In Ladd, H. F., and Fiske, E. B. (Eds.), *Handbook of research in education finance and policy.* New York: Routledge.

Education Week (2012). *Quality counts, 31*(16), 62.

Hanushek, E. (September 1986). The economics of schooling: Production and efficiency in public schools. *Journal of Economic Literature, 24,* 1141–77.

Hussar, W., and Sonnenberg, W. (2000). *Trends in disparities in school district level expenditures per pupil.* Washington, DC: National Center for Education Statistics. Retrieved from http://nces.gov/pubsearch/pubsinfo.asp?/pubbid=2000020

Johns, R. E., Morphet, E. L., and Alexander, K. (1983). *The economics and financing of education* (4th ed.). Englewood Cliffs, NJ: Prentice Hall.

Mort, P. R. (1924). *The measurement of educational need.* New York: Teachers College Press.

O'Day, J. A., and Smith, M. (1993). Systemic school reform and educational opportunity. In Fuheman, S. (Ed.), *Designing coherent educational policy: Improving the system* (pp. 250–311). San Francisco: Jossey-Bass.

Ramirez, A. (2003). The shifting sands of school finance. *Educational Leadership*, *60*(4), 54–57.

Strayer, G. D., and Haig, R. M. (1923). *The financing of education in the state of New York.* New York: MacMillan.

Updegraff, H. (1922). *Rural school survey of New York State.* New York: Author.

State School Finance Systems   **6**

Aim of the Chapter

THIS CHAPTER IS DESIGNED TO HELP THE reader understand how states typically allocate funds for the support of schools. The fundamental components, commonly found in state school finance formulas, are examined along with theories of school finance. Guiding principles for good school finance policy are also presented.

Introduction

The fifty states and territories distribute a combined $600 billion annually for the operation of the 15,000 school districts and 100,000 schools across the United States. Revenues for these annual appropriations come from such sources as income tax, corporate tax, sales tax, property tax and lottery proceeds, among many others (NCES, 2010). Each state has devised its own unique method of raising and distributing funds.

These state "school funding formulas" have developed over time within each state in a dynamic environment affected by politics, economics and litigation. Given these complex circumstances, it is easy to understand why most reactions to discussions about school funding formulas quickly elicit the glassy-eyed stare of confusion or boredom. This chapter uses basic information and accessible language to explain how it happens.

The Authority to Fund

The power to raise money for schools is found in the state constitution, which declares the intent and obligation of the state to operate a system of schools. Such state constitutional articles or amendments either implicitly or explicitly provide for a means of funding schools. The state legislature, based on this constitutional mandate, devises a method for collecting money and supporting the operation of the schools through statutes, which detail how money will be collected and distributed.

Together with the attendant regulations, the constitution and laws work to achieve several overarching goals, which are common to most state school funding formulas. These goals include:

o To offer an equal educational opportunity to all students;
o To raise sufficient money to operate the schools in a manner that enables school districts to reach student outcomes desired by the state;
o To maintain a fair distribution of resources between school districts;
o To balance the burden of raising money for schools among local communities, and between local communities and the state;
o To set parameters—upper and lower limits—on how much tax money must be collected and can be collected for the support of schools;
o To set limits on the size of school district budgets;
o To ensure each school district offers the programs and serves the students designated by the state;
o To be sensitive to the unique circumstances found in various communities across the state, such as student demographics, geographic setting, economic conditions and school district size.

Each state crafts a unique funding mechanism to accomplish these goals, although some components are common to many state formulas. While the goals of a state funding system will hold fairly steady, over time these formulas will change because of such things as litigation or the threat of litigation, shifts in the tax structure of a

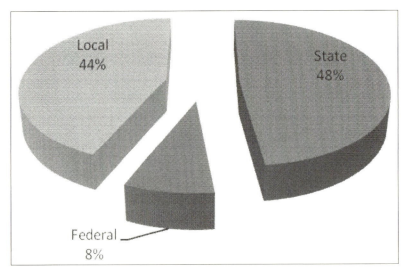

Figure 6.1   Averages of revenue sources for PK–12 education across the nation.
Source: Digest of Education Statistics, Washington, DC: National Center for Education Statistics.

state, changes in the economic conditions in the state, demographic changes that shift dominant voting blocs from rural to urban to suburban areas, new thinking about school finance theory or shifts in political power.

Meeting Funding Goals

How much do you need? How much do you have? These two questions lie at the heart of the annual struggle to determine the level of funding for schools. There are examples of states setting funding appropriations for more than one year at a time, but even in these situations the legislature must consider such things as revenue collections and changes in enrollment on a yearly basis. This applies even in states like Montana, Nevada and Texas, which have biennial legislative sessions.

The process of deciding how much schools should get each year is one of the most important acts of the state legislature. The budget for schools is often the largest single appropriation in the state budget. Therefore, decisions about school funding affect many other areas of the state budget. Additionally, it is important to remember that every legislator has a stake in the outcome of the education budget decision, since every legislator has schools within his or her legislative district.

While major elements of funding formulas typically change little during the life cycle of the formula, important but subtle, often seemingly minor changes and "adjustments" are watched closely by the various political players, interest groups and school stakeholders. A medium-size school district will have an annual operating budget of over $100 million and the largest districts easily reach $1 billion or more. Slight changes in a state formula can translate into millions of dollars, which can be a boom or disaster for an individual district. So, the state funding formula is very important and watched very closely.

How Much?

How does a state determine how much money it should spend on the schools for a particular year? Numerous variables are considered as the state legislature, governor, state board of education, chief state school officers and a host of interest groups contemplate this question. If there is a high degree of confidence in the state funding formula then that will serve as the main focus for calculating the amount. If, on the other hand, there is much dissatisfaction with the current formula, efforts to establish a new formula or make major revisions to the formula will be the focus of the legislative debate.

It is not uncommon to see the legislature create funding streams, or programs, both inside and outside the formula in order to achieve a political consensus and get the school funding bill approved for yet another year. This is done in two ways:

1. creating categorical programs that direct funding to selected school districts outside the funding formula;
2. using "weights" when counting students for appropriation purposes.

Categorical programs often fund specified groups of students, for example, new learners of English, students living in poverty, gifted students or preschool children. Categorical programs can also be based on school district characteristics, such as school districts in remote rural areas, school districts with high proportions of public lands, or school districts with unique economic conditions, like the sudden loss of a major industry.

Weights are used to direct more funding to school districts with higher percentages of certain kinds of students, for example, students with disabilities, students who are poor or new learners of English. In such cases different classifications of students will count more or less than a full-time equivalent student for apportionment purposes. Weights work by adding value to the amount of money an individual student generates in the formula and are based on the student's educational need. As an example, an "average" student might generate 100 percent of the base state per pupil allocation, a new learner of English 125 percent and a student with an individual education program 140 percent.

Categorical programs and weighted enrollments modify the distribution of funds based on the unique needs of students and school districts. According to a study by the Education Commission of the States, twenty-five states use weights in their formula, twelve states use various forms of categorical programs and thirteen states use a combination of both with adjustments (Griffith, 2005).

Who Gets What?

Never is the aphorism that "politics is the business of deciding who gets what" more apparent than when a state decides how much money should go to the schools and how much each school district will receive. It is naïve to believe that school finance formulas are totally rational and objective creations. They exist within a rational and objective framework, but ultimately the realities of available revenue and politics decide how much money goes to which school districts. History is a good teacher in this regard.

Consider how the population has shifted within most states over the past one hundred years from rural communities, to the big cities, and then to the suburbs. These shifts in population are reflected in the balance of power within the various state legislatures and in the distribution of funding for schools. A legislature that historically was supportive of hundreds of small school districts now emphasizes the consolidation of school districts into larger, "more efficient units." Years later the legislature, through the state funding formula, emphasizes funding for disadvantaged children, who tend to be concentrated in the big cities. In recent decades, the trend has been to underfund categorical programs that serve special populations and put available new dollars into the state funding formula in a way that gives more dollars to suburban school districts, which in many states tend to have more powerful representation in the state legislature (Ramirez, 2003).

Ultimately, how much is appropriated for education funding in a state is based on what is available to spend, which is based on the condition of current or projected revenues, which is based on the economic cycle and political decisions about where to spend public monies. Questions of equity and adequacy (discussed in chapter 5) are ongoing debates shaped by politics, prevailing public opinion, research and the courts.

Guiding Principles for State Funding Formulas

As was stated above, a state school funding formula must exist within a framework that is rational and objective. Politics aside, the formula must be relatively fair, at least to most stakeholders. Two reasons for this are that it must successfully complete the legislative process, including the signature of the governor, and it must withstand the threat of litigation. A funding scheme that is biased against a classification of students is subject to being thrown out by the courts. Generally, legislatures hate the idea of the courts telling them how they should legislate. Thus, individual legislators often try to stretch the boundaries of their political power by favoring their constituents at the expense of others, while not becoming so egregious as to become susceptible to the attention of the courts.

Despite all the politics associated with school finance, there are a number of guiding principles that can be applied to the development or evaluation of a state school finance system. These principles should serve as universal criteria against which a funding formula can be judged. A number of widely accepted guiding principles have developed over time based on evolving theories of school finance, better technology and information, and emerging legal doctrines related to education and funding. Augenblick, Fulton and Pipho (1991) articulated a set of guiding principles for good school finance formulas that have held up over time:

1. The formula and financing system are fair.
2. All school districts are considered.
3. The amount of funding generated by the formula is sufficient to achieve desired educational results for all students.
4. All needs are considered, e.g., transportation, teacher recruitment, facilities maintenance.
5. The unique circumstances of individual school districts are factored into the formula, for example, school district size, population density or sparseness, concentrations of poverty, economic conditions.
6. Differences in per pupil funding between school districts are reasonably explained.
7. The tax burden within the school district is neither excessive nor too low and is linked to the wealth of the school district.
8. School district wealth is measured in several ways, not just property.

9. The state levels up funding for those school districts that cannot meet targeted funding amounts locally.
10. School districts have an option to raise money, on a limited basis, outside the funding formula, exclusively with local sources.
11. Local control is maintained within a state system.
12. Funding is reasonably predictable, both short and long range.

The Foundation System

It was in the early part of the twentieth century that scholars, school administrators and political leaders gained support for the concept of providing additional assistance to school districts that were incapable of raising sufficient local revenue to provide for an adequate education. Recall that funding for schools in the nineteenth century and earlier was overwhelmingly a local responsibility and involved only minimal state-level funding. During this earlier time, state funding tended to be structured to provide incentives to local communities, for example, to start schools, expand services to unserved populations or move school districts toward the standardized system (equated to quality at the time) of the public schools envisioned by the state.

During the latter half of the nineteenth century and the first half of the twentieth century, the states were focused on building the capacity of the public school system to meet the needs of the burgeoning number of students. Concurrent with this thrust was the movement to have all school districts and schools meet state standards, typically related to uniformity of the design or inputs of the system. For example, the number of days in the school year, the length of the school day, the course of study for elementary and secondary schools, the layout of school buildings and the qualifications of teachers and other school personnel were the focus of quality improvements by the state.

It was also common during the nascent era of the public schools, on those occasions when the state did provide funds, for the state to use lump sum grants and flat grants as the mechanism for allocating revenue. Lump sum grants, as the name implies, were given out to communities in equal amounts. That is, each city, town or village was provided the same amount of money, without regard to number of students, teachers or schools. Flat grants were seen as an innovative development, which aimed to distribute funding based on the number of students, teachers, schools or other similar unit of measure. Thus, size and proportion were considered in the decision to allocate flat grant funding.

Studies conducted by pioneers like Ellwood P. Cubberley, George Strayer, Robert M. Haig, Harlan Updegraff and Paul R. Mort in the early part of the twentieth century revealed the deficiencies of the system primarily based on using local tax revenue to fund schools (Johns, Morphet and Alexander, 1983). The research demonstrated that although each community had equal authority to levy taxes locally, the resulting amounts raised varied greatly. Furthermore, the researchers were able to show how the defective funding system led to large variations in the quality of education from community to community. Based on the standards of the time, policy analysis revealed that some students could not get an adequate education because the community could not fund their schools properly.

The concept of the *foundation* system was created to overcome problems of deficiency in funding due to unequal wealth distribution among school districts. The theory behind the foundation approach rests on several concepts:

1. every student is deserving of a minimally adequate education;
2. the cost of a minimally adequate education is knowable;
3. the state has an obligation to help fund the education of children who reside in communities that cannot raise what is needed locally.

To this day these concepts influence thinking about school finance.

The foundation program works through the use of mechanisms, usually embedded in the state funding formula, targeted toward ensuring that each student has a *guarantee* on the amount of money supporting his or her education. This guarantee is the foundation, or floor, upon which education funding is built. The guarantee, or foundation amount, is supposed to represent the amount of money needed to achieve a basic, adequate or minimal education for a theoretical individual student. How "adequate," "basic," "minimal" or other foundational terms are defined is a matter undertaken by each state and often debated in the chambers of the state legislatures and the state courts.

Under the foundation system each school district makes a good faith effort to raise sufficient money to at least reach the guarantee (or foundation) level. School districts that cannot raise the minimum foundation level on a per student basis get state support to help them reach the target amount per student. The examples below illustrate the system.

Scenario 1

Imagine a mythical state called Franklin. The legislature in Franklin has determined that in order for a typical student to receive a quality education, the student must have $10,000 supporting his or her education each year. Thus, the legislature sets $10,000 as the "minimum guarantee" amount per pupil in the state. In the state of Franklin, school districts must raise revenue for schools through local property tax.

School districts A and B (Aspen and Birch) are about the same size, and both levy taxes on local property at the same rate, as specified by state law. The Aspen school district is in a high-income area with big homes and commercial properties that are assessed (valued) well above the state average. The Birch school district, on the other hand, serves a more modest community and property values are well below the state average. Through the required taxation, the Aspen school district easily reaches the $10,000 per pupil target, and in fact exceeds the target by 40 percent. But the Birch school district can raise only $8,000 per student

Picture 6.1   The amount spent on a child's education should not be determined by the property wealth of the community.

and thus misses the target by 20 percent. Under the foundation plan, the state, through other revenue sources, would "level up" the amount per student for the Birch district to help it reach the $10,000 per pupil guarantee, in this example, $2,000 per pupil.

Scenario 2

Continuing our example with the Aspen and Birch school districts, we see in this scenario the Aspen school district has a property tax base, or total assessed valuation, that is so large that when it applies the required state tax rate (i.e., levy), it actually raises $14,000 per pupil. This is $4,000 beyond the state guarantee per pupil of $10,000. Thus emerges a common set of dilemmas encountered by states. Should the state allow Aspen to keep the extra money? Should it require Aspen to lower its property tax rate and thus collect less revenue, but just until it reaches $10,000 per pupil? Should it require the Birch school district to tax itself at a higher rate to get closer to Aspen? Should it take some money from Aspen and give it to Birch? Each option has its own set of nettlesome issues.

Scenario 3

If Aspen is allowed to keep the extra money it can easily raise, the state will quickly run into a problem related to *equalization* of funding across the state.

Picture 6.2   The amount spent on a child's education should not be determined by the property wealth of the community.

Equalization addresses the concept that each child, within explainable and justifiable differences, should have about the same amount spent on his or her education, regardless of where the student lives in the state. It further underscores the concept that a child's education should not be diminished because of where the child lives in the state. In other words, the state has a constitutional obligation to treat all students equally. Allowing Aspen to keep its extra money creates a policy that leads to a dual funding system in which property-rich communities spend substantially more per pupil than property-poor school districts. The courts have ruled in many states that this is not legal, because it violates equal protection provisions in the state constitution.

Scenario 4

An alternative solution is to lower property tax rates in the Aspen district to the level that the district only raises an amount equal to the state $10,000 guarantee. While this would help address the question of equalization of funding between school districts, it moves the problem into a new realm, the issue of fairness for taxpayers. Lowering the tax rate for Aspen residents creates unfairness for the Birch residents, in that they must assume a higher tax burden relative to their wealthy neighbor. Thus, lowering the tax rate for wealthy districts creates a dual system of taxation, one for property-rich school districts and one for property-poor school districts. Equalization must also apply to taxpayers across the state, so that all individuals are taxed fairly for government services.

Scenario 5

Opting to take some funding from the Aspen school district and transfer it to the Birch school district is yet another possible solution. This "Robin Hood" approach, taking from the rich to give to the poor, has been very contentious in states where it has been tried. Property owners are loath to pay for someone else's government services. Consider how you would you feel about being taxed to pay for the fire department in a neighboring town. Some states have managed to overcome this hurdle by establishing a statewide property tax and redistributing those revenues for equalization purposes.

Legislators, governors and education policy makers at all levels have extensive and ongoing challenges to establish systems of funding that are adequate to both support established educational goals and maintain fairness for their citizens. Through an ongoing evolution of state school funding systems, several new methods are being used or considered to meet these challenges. Here are some examples:

o More state aid and less reliance on local property tax.
o Expanded revenue sources for school districts, such as sales, gaming and payroll taxes.
o Realignment of the state taxing structure to capture and distribute more funding from the state; for example, taxes on some commercial real estate and utilities would flow directly to the state.
o The state assumes debt related to some aspects of the school district like construction costs.
o Categorical programs are used to equalize funding by having the state pay for most or all of such programs, e.g., special education, transportation and vocational education.
o Pool districts within a taxing region to equalize revenue within the region.

School finance theorists such as John Augenblick and Alan Hickrod have added significantly to the knowledge of the past and help current policy leaders develop ever more sophisticated school funding systems. They have helped policy leaders develop funding formulas that more closely match the ideals and principles of fair and adequate systems of school funding.

One trend that resulted from the struggle to devise funding mechanisms that meet the criteria for a fair and adequate system has been an increase in state aid to the public schools. During the early development of public education funding was primarily a local matter. Over time this has shifted so that today, on average across the nation, the state is the major source of school aid.

Table 6.1   Financing sources for state funding systems, state provided formula funds, state reclaim of funds and funding systems ruled unconstitutional, by state: 2004–05.

State	Financing sources for state funding system[1]	State requires a minimum local effort for districts to receive state aid[1]	State reclaims funds from districts able to generate above a specified amount[1]	Current state funding system has been ruled uncon-stitutional for equity concerns (2003–04)[2]
United States	†	35[3]	5[3]	9[3]
Alabama	Foundation	Yes	No	Yes
Alaska	Foundation	Yes	No	No
Arizona	Foundation	Yes[4]	No	Yes[3]
Arkansas	Foundation	Yes[4]	No	Yes
California	Foundation[6]	No	No	No
Colorado	Foundation	Yes[4]	No	No
Connecticut	Foundation	Yes	No	No
Delaware	Flat grant/local-effort equalization	No	No	No
District of Columbia	Foundation	†[7]	†[7]	No
Florida	Foundation	Yes	No	No
Georgia	Flat grant/local-effort equalization	Yes	No	No
Hawaii	Full state funding[8]	†[7]	†[7]	No
Idaho	Foundation	Yes[4]	No	No
Illinois	Foundation/flat grant	Yes[4]	No	No
Indiana	Foundation[9]	Yes[4]	No	No
Iowa	Flat grant/local-effort equal-ization	Yes	No	No
Kansas	Flat grant/local-effort equal-ization	Yes	Yes	No
Kentucky	Flat grant/local-effort equal-ization	Yes	No	No
Louisiana	Flat grant/local-effort equal-ization	No	No	No
Maine	Foundation	Yes	No	No
Maryland	Flat grant/local-effort equal-ization	Yes	No	No
Massachusetts	Foundation	Yes	No	No
Michigan	Foundation	Yes[4]	No	No
Minnesota	Flat grant/local-effort equal-ization	No	No	No
Mississippi	Foundation	Yes	No	No
Missouri	Foundation[9]	Yes	No	No
Montana	Foundation	No	No	No
Nebraska	Foundation	Yes[4]	No	No
Nevada	Foundation	Yes	No	No
New Hampshire	Foundation	No	No	No

New Jersey	Foundation	Yes	No	Yes
New Mexico	Foundation	Yes	No	Yes[5]
New York	General aid[10]	No	No	Yes
North Carolina	Foundation	No	No	Yes
North Dakota	Foundation	Yes[4]	No	No
Ohio	Foundation	Yes	No	Yes
Oklahoma	Foundation	No	No	No
Oregon	Foundation	Yes[4]	No	No
Pennsylvania	Percentage equalization[11]	No	No	No
Rhode Island	General aid[12]	No	No	No
South Carolina	Foundation	Yes	No	No
South Dakota	Foundation	Yes[4]	No	No
Tennessee	Foundation	Yes	No	No
Texas	Foundation/local effort equalization	Yes	Yes	No
Utah	Foundation	Yes	Yes	No
Vermont	Full state funding	No	No	No
Virginia	Foundation	Yes	No	No
Washington	Full state funding/local-effort equalization	No	No	No
West Virginia	Foundation	Yes	No	No
Wisconsin	Guaranteed tax base	No	Yes[13]	No
Wyoming	Foundation	Yes	Yes	Yes[5]

Source: Education Week, Quality counts 2005, table Resources: Equity. http://nces.ed.gov/programs/statereform/source.asp#education.

† Not applicable

[1] *Education Week*, Research center annual state policy survey, 2004.

[2] *Education Week*, Quality counts, 2004.

[3] United States total number of affirmative or "Yes" responses for each column.

[4] A minimum local effort is not required for districts to receive state aid; instead, the state assumes local districts will raise a certain amount and adjusts state aid accordingly.

[5] Rulings for these states were based on funding for school construction.

[6] California has several grants and entitlements in its school funding formula, the largest of which is general-purpose aid. General-purpose funding is based on a modified foundation formula, and the foundation level varies for each local education agency.

[7] Hawaii and the District of Columbia both are single school districts.

[8] Hawaii basis of state funding formula based on the 2003–04 school year.

[9] Indiana's school finance system is based on a foundation program, but the state uses a guaranteed tax base formula to determine the local share. Missouri calculates its foundation level by multiplying a guaranteed tax base by a minimum required tax rate.

[10] The combination foundation/percentage equalizing formula that generated operating aid in New York state for many years has not been used as the basis for allocation of that aid since the 2000–01 school year. For 2004–05, every district received a 1.75% increase from its 2003–04 funding level.

[11] In Pennsylvania, the subsidy from the prior year has been the base for the current year; any additional funding for the current year has been distributed through various formula components called supplements. The base supplement is based on a district-wealth ratio.

[12] Rhode Island uses 10 major methods to distribute education funds. The largest dollar amount, general aid, is a fixed amount based on what a district received in fiscal year 1998.

[13] There is recapture in Wisconsin if a school district has "negative aid" in Wisconsin's third tier of funding. Although funds are not returned to the state, those districts share local funds with districts that have property wealth lower than the state average.

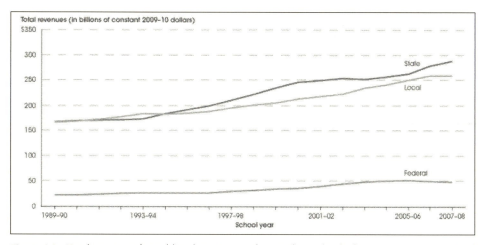

Figure 6.2   Total revenues for public elementary and secondary schools, by revenue source: School years 1989–90 through 2007–08.
Note: Revenues are in constant 2009–10 dollars, adjusted using the consumer price index (CPI). For more information about the CPI and revenues for public elementary and secondary schools, see *supplemental note 10*. For more information about the Common Core of Data, see *supplemental note 3*.
Source: U.S. Department of Education, National Center for Education Statistics, Common Core of Data (CCD), *National Public Education Financial Survey*, 1989–90 through 2007–08.

Variations of the foundation approach have been tried and applied by manipulating the revenue side of the funding equation, for example, guaranteed tax base, guaranteed yield and power equalization. A *guaranteed tax base* uses the tactic of treating each school district as if it had a tax base sufficient to reach the desired target amount per pupil. A specified tax is applied by all school districts. For those school districts that do not raise the specified amount, the state makes up the difference between what tax revenue is actually produced and the target. The *guaranteed yield* is another such method. This technique assumes a target "yield" per unit of taxation. Districts that fall short of the target get state aid. A third procedure is the *power equalization* scheme. Under this setup districts whose revenue exceeds the state guarantee amount would see some or all of these excesses collected by the state for redistribution to poorer school districts—an obviously contentious maneuver. Permutations of these *reward for effort* designs are seen in state funding systems around the country (Griffith, 2005).

Not surprisingly, some states will incorporate a combination of these methods to achieve desired balances across their state. The use of tiers allows the state to add different elements to its funding formula. In such cases when a tier, or level, is met, another funding approach is added. For example, a school district must tax itself at some state-established rate in order to participate in the equalization program. Another example might allow a school district to collect additional local revenue beyond the guarantee, but this amount is not equalized. Table 6.1 shows how states have mixed formulas.

A big question associated with funding schools is whether a school district has the ability to pay. The ability to pay is based on the wealth of the community. However, how wealth is determined is very significant. States commonly use assessed valuation of real property as a measure of school district wealth. The *ad valorem*, i.e., value-added, property tax is a good source of tax revenue because it is fairly easy to determine and historically has tended not to fluctuate as drastically with the business cycle as other revenue sources such as income or sales tax. However, recent declines in the value of real property across the nation caused by the "housing bubble" have unsettled this historically stable revenue source. How a state chooses to tax property and how it classifies property can affect a school district's revenue and thus its ability to pay.

Residential, commercial, industrial, utilities, agricultural, mining and forests each have their own unique economic value. School districts may have more or less of each such property type within their boundaries. The state's method of assessing these properties and where it assigns the revenue stream from the tax can help or hurt the financial position of a school district. Having sufficient sources of revenue is an important consideration as the state ponders the school funding system.

While assessed property value is a good indicator of a school district's capacity to pay, it is appropriate to consider other sources of wealth. Per capita income is an example. This is particularly true when the mix of real property in the school district is limited, say, to rental property, which may be taxed at

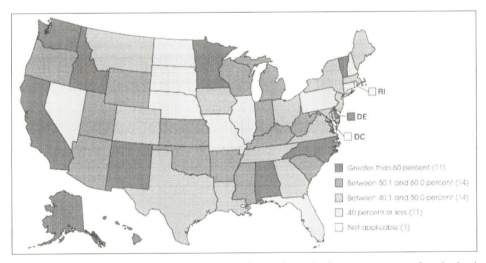

Figure 6.3   State revenues for public elementary and secondary schools as a percentage of total school revenues, by state: School year 2007–08.

Note: Both the District of Columbia and Hawaii have only one school district each; therefore, neither is comparable to the other states. For more information about revenues for public elementary and secondary schools, see *supplemental note 10.* For more information about the Common Core of Data, see *supplemental note 3.*

Source: U.S. Department of Education, National Center for Education Statistics, Common Core of Data (CCD), *National Public Education Financial Survey,* 2007–08.

a lower rate than other forms of property. Sales tax is yet another example for school districts that have large commercial centers like shopping malls. Determining a school district's ability to pay requires an understanding of measures of wealth and judgment about the effort made by the school district to collect required funding.

In contrast with school district wealth is school district need. Circumstances related to everything from geographic location to demographic makeup cause school districts to have different financial needs. Here are some examples:

Table 6.2   The development of school finance systems.

State aid as a portion of school district revenue	Timeline	
	2000	Adequacy lawsuits launch new era of increased state support to school districts. New criteria developed to judge state funding formulas based on student outcomes. Increased spending from the state level to fund categorical programs for special populations. State aid to school districts represents the major portions of funding in many states.
50% +		
	1980	States experiment with variations of foundation system. Reward for effort and multi-tiered funding schemes introduced. Power equalization programs and guaranteed revenue support tried. Local property taxes contribute smaller portion of school support in many states.
40%		
30%	**1960**	Funding equity lawsuits usher in new era of state support for school districts. Use of categorical programs based on the classifications of students, e.g., poor and disabled. Incentives grant funding from state and federal level target curricular areas.
16.5%		Foundation programs adopted in states around the country. States guarantee a base amount of funding per pupil regardless of local wealth.
	1920	
		Era of flat grants to school districts based on enrollment, teacher units or number of schools.

o A school district with a large geographic area through which many small communities exist may have to maintain numerous small attendance centers, whereas a school district of similar enrollment contained in a compact geographic area can build fewer, more efficient schools.

o Two school districts of equal enrollment, but one has an immigrant population of 30 percent and must offer extensive English language acquisition programs.

o A school district with rapidly declining enrollment or a school district with rapidly expanding enrollment.

o School districts with high percentages of low-income families.

o School districts with inordinate transportation costs.

School funding formulas should account for differences among the school districts in the state. The idea that equity is treating all districts equally when distributing resources, without consideration of the characteristics of the school district, just does not hold up. An equitable funding formula takes into account both the wealth and the needs of the school district.

Summary

School funding formulas are what states use to collect and distribute money for the operation of schools. These funding systems have evolved over time as research, politics and litigation have come to bear. State finance programs must strive to be fair to all stakeholders from students, to teachers, to taxpayers. One result of the effort to achieve equity in these systems is that the proportion of state aid compared to local support has increased substantially in many states.

Property taxes are a common revenue source for school districts and are generally reliable and stable. However, overreliance on property taxes can cause disparities between property-rich school districts and property-poor school districts. New sources of revenue and new means of distributing revenue are under constant review. Accounting for a school district's ability to pay and its unique needs are also important aspects of a state funding formula.

Chapter 7 looks at several of the alternatives to funding education being used or advocated in a number of states. These approaches differ significantly from the state funding systems used to support school districts. School vouchers and charter schools are among the methods considered.

References

Augenblick, J., Fulton, M., and Pipho, C. (1991). *School finance: A primer*. Denver, CO: Education Commission of the States.

Griffith, M. (2005). *State education funding formulas and grade weighting*. Denver, CO: Education Commission of the States.

Johns, R. E., Morphet, E. L., and Alexander, K. (1983). *The economics and financing of education* (4th ed.). Englewood Cliffs, NJ: Prentice Hall.

National Center for Education Statistics (2010). *Digest of education statistics.* Washington, DC: National Center for Education Statistics.

Ramirez, A. (2003). The shifting sands of school finance. *Educational Leadership, 60*(4), 54–57.

Alternative Funding Systems

7

T HE PURPOSE OF THIS CHAPTER IS TO PROVIDE insight into the theory of school choice and understand why advocates propose this policy as a vehicle for equity and efficiency in attaining an educated population for the nation. The chapter also aims to develop an understanding of the scope and variety of alternative funding programs and education options existing within school choice policy. The fiscal implications of the various school choice approaches are explored, as well as an understanding of how the concept of "the child benefit" theory within these alternative funding schemes is applied.

Introduction

The system of tax-supported public schools in America, available to all children, developed over many decades. As recounted in chapter 1, the public schools in place today came into existence after many long policy and political battles. Throughout the extensive period of advancement, advocates for and against the concept of tax-supported common schools debated vigorously for their respective positions. Numerous and varied constituencies comprised all sides of the battle line.

Today, two centuries later, we see many of the same arguments proffered by the same groups, new groups with old arguments and old constituencies with new arguments, and there is a continuing battle over how best to attain the goal of an educated citizenry. Just like in the eighteenth century, there are those who say education is a private matter and solely the responsibility of the child's parents. Others argue that education is so important to the economic and civic viability of the nation that universal education must be provided by the state. Some assert education is mostly a private benefit and thus should be funded by the individual or the family; others emphasize the benefit to society of an educated individual. Issues of social justice, social cohesion and personal well-being are in the mix as well.

But it is beyond the scope of this chapter to delve into the nuances of each economic, political, cultural and legal line of reasoning. Instead, this chapter investigates the various methods in existence today for funding PK–12 education, which operates outside the conventional public school system. The theory of action for each method is explained along with some basic information about the scope of the program and its effect on public school funding.

Topics explored in the chapter include:

o school choice as a theory, policy and practice;
o open enrollment programs;
o school vouchers, including state aid to private school students, private school tuition tax credits, private school tuition reimbursements and tax-funded scholarships for private school students;
o state aid to private schools through supplemental aid, categorical funding and tax-funded endowments to private schools;
o charter schools;
o home school or home education;
o and funding options used in other countries.

School Choice Theory

The legal foundation for the rights of parents to direct the education of their children is well-settled law in the United States. The U.S. Supreme Court has consistently held that as long as a child is being educated in some basic way, the state cannot impose overbearing policies as to the form or manner of education. In 1925, *Pierce v. Society of Sisters* established that parents could send their children to private schools in the face of an Oregon law that mandated public school atten-

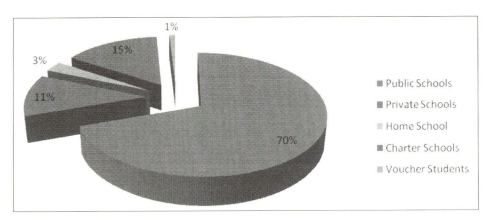

Figure 7.1   Percent of enrollment by choice options across the United States.
Calculated from the *Digest of Education Statistics* (2010). Washington, DC: National Center for Education Statistics.

dance for all. *Wisconsin v. Yoder* in 1972, on the other hand, addressed the issue of how much education could be required by the state and helped further define the parameters of where the state's interests and parent's interests lie. Whereas by 2002, *Zelman v. Simmons-Harris* circumscribed the conditions under which tax-supported voucher funds could be used in private religious schools, thus illuminating the distinction between state aid for the support of religion and state aid for the support of a child's education.

In its simplest form, school choice is a policy theory that advocates for the ability of parents to select their child's school or manner of education. However, the policy theory has no authoritative definition and so is used as an overarching label, which covers a multitude of situations. For example, school choice policies can range from a completely privatized system of elementary and secondary schools from which parents may select any school to a policy that allows parents to select among public schools within their school district of residence. Many policy and program configurations exist, or are proposed, within these extremes.

The theory of action for school choice policies stems from three broad principles.

1. One principle is grounded in economic theory and views school choice models as an essential mechanism for improving the PK–12 education system by fostering competition among schools. This economic principle envisions a marketplace dynamism in which schools compete for student enrollment. The theory proffers that "good" schools will thrive and "bad" schools will go out of business as parents make choices in the competitive market (Friedman and Friedman, 1980).

2. A second theory of action within school choice policy thinking is based on principles and legal thinking related to liberty. This theory asserts that the ability to choose one's school is a fundamental right. Advocates of this view argue that it is wrong to sustain a system of education that is monopolistic in nature, offers no choices to parents and requires a child to attend one school designated by his or her school district. Choice in this model supports parental rights to move a child from a failing school or seek a school with a curriculum that better suits the family. This policy perspective casts the school choice debate in terms of freedom from an overbearing government.

3. A third theory of action associated with school choice, and one that is at the heart of school finance principles in this arena, is the child benefit theory. This concept asserts that public money flowing to a school—public, charter or private—is not sent for the benefit of the school, but rather to benefit the child attending the school. The child benefit theory links to compulsory education policy, which addresses the societal interest in having all children educated to some functional level. Choice policy theory supports the idea that education can take place in many settings, which can serve the societal interests of an educated nation.

Picture 7.1   Many families seek alternatives to the neighborhood public school.

Because school choice is such a broad policy concept, it attracts supporters from across the political spectrum. Therefore, some policy camps see school choice as a matter of basic liberty, some see it as a matter of economic efficiency and still others perceive school choice as a question of social justice. The theories of action presented above are offered as background information to help explain some of the policy goals behind the choice options presented in this chapter.

Open Enrollment Policies

This policy concept is the simplest form of school choice. Within this structure parents are free to choose any public school for their child to attend. In some states this choice is restricted to schools within their school district of residence; however, other states allow choice between school districts as well. Exceptions to this policy exclude some magnet schools or schools established for children with certain disabilities from the choice universe. But, under an open enrollment policy, parents are free to pick from among all public schools.

The pattern of building "regular schools" is typically based on the idea that the school should be where the children are, i.e., near the child's home. Efficiency is one of the key factors that drives the decision about where to build a school. Minimizing the cost of transporting students to schools distant from their homes is an example of such an efficiency concern. In some cases the open enrollment policy will come with transportation. In other cases the transportation is limited to the one school nearest the student's home, so parents who choose other schools must provide transportation. Another variation seen with inter-district choice sees the home district providing transportation within district, but not across district borders. Depending on how the policy is structured, open enrollment can be more or less costly for the district.

When local school boards lose the authority to assign students to schools because of state or federal policy, other cost factors come up. With open enrollment, some schools can be underutilized and others oversubscribed. This is consistent with the intent of the policy. School districts are then forced to shift resources around to match needs at individual schools. For example, they must add programs to under-enrolled schools to make them more attractive to school choosers, and send more resources to popular schools to match demand. Here again, efficiency in the use of resources is sacrificed.

School Vouchers

At the other end of the school choice spectrum is the policy that provides vouchers for educational expenses. School vouchers are not new in the United States and are found in countries around the globe like Italy, Denmark, Chile, and Hong Kong. However, the programmatic manifestation of this policy is seen on a limited basis in the United States. While states like Vermont and Maine have long used a form of vouchers to allow children in towns without secondary schools, for example, to attend such schools in neighboring towns, voucher programs in other states are limited and experimental.

Within school choice policy, a voucher is a guarantee of payment by the government for enrolling a student. In the program's most unencumbered theoretical form, all children would be eligible for a voucher and be able to enroll in any school, public or private. This pure form of the voucher program does not exist in the United States, as most extant voucher policies found in the states have eligibility requirements for schools and students, as well as quotas, limitations and restrictions.

Seventeen states and the District of Columbia currently have some form of voucher or quasi-voucher program. What is seen today in the United States within the realm of voucher policy are more limited programs designated for particular school districts, like in Milwaukee, Wisconsin; Cleveland, Ohio; and Washington, D.C. Most programs are limited to certain populations—like poor children or children with disabilities—such as in the states of Arizona and Florida. These programs offer a fixed amount per child and prohibit the accepting school from charging additional tuition above the voucher amount. In those cases where tuition is below the voucher amount, the government also prohibits the accepting school from charging more than its stated tuition for voucher students. Most programs today cap the number of students who may receive a voucher and resort to lotteries for the distribution of vouchers.

The amounts offered in existing voucher programs are typically less than the per pupil amount spent in the child's school district. From a cost perspective, this saves the state and local taxpayer a portion of the cost of educating the child had the parent decided to enroll the child in the public school. This provides a financial incentive for the state to entice families to accept vouchers. Transportation and other supplemental services are not typically covered, so those costs are shifted to the family. Thus, the implications for children from poor, middle-class and wealthy families are different within a voucher system.

Variations of the school voucher program found among the states include state aid to private school students, private school tuition tax credits for families with children in private schools, private school tuition reimbursements and tax-funded scholarships for private school students. Vouchers typically go from the government to the family to the school (public or private) and back to the government for payment. The scholarship process moves money from the taxpayer (corporate or private) to the scholarship organization to the parent to the school. Reimbursement programs flow from the government to the family, most often in the form of a tax credit against personal income tax.

Arizona, for example, allows corporations to contribute to a private school scholarship organization and receive a tax credit for their contribution. In Iowa families earn a tax credit on their state income tax for a small portion of the cost of sending a child to private school. In Vermont a sending town pays the receiving town a per pupil amount. Participation rates and dollar amounts vary greatly among the states and the programs. For example, the Ohio Autism Program will provide $20,000 per student for a private school placement, but the Iowa tax credit is capped at $250 for a child or $500 per family for private school expenses. Some state programs have small participation while others serve thousands of students. Table 7.1 lists the array of voucher-related programs extant in the nation.

Several of the states with petition initiative ballot mechanisms, such as California, Colorado and Utah, have held plebiscites on the question of school vouchers. When presented to the voters throughout the state, vouchers tend to be overwhelmingly voted down. This is particularly true for universal voucher programs open to all PK–12 students. Therefore, most of the existing voucher programs have been created by the legislature, limited to certain populations and capped in terms of the number of students or total funding allocated to vouchers. Theoretically, some of the "scholarship" programs that offer tax credits to individual or corporate donors could provide unlimited vouchers to families seeking such support, but this has not happened.

State Aid to Private Schools

About 9 to 11 percent of the school-age population in America attends private schools. This percentage has been fairly consistent since the 1970s, although declines have been noted as the popularity of charter schools has taken hold. According to the *Digest of Education Statistics* (NCES, 2010) there were 33,740 private

Table 7.1   Voucher-type programs in the United States.

State	Program Type	Name of Program
AZ	Scholarship	Corporate Tax Credits for School Tuition Organizations
AZ	Gov. Grant	Displaced Pupils Choice Grants
AZ	Scholarship	Personal Tax Credits for School Tuition Organizations
AZ	Voucher	Scholarships for Pupils with Disabilities
DC	Voucher	Opportunity Scholarship Program
FL	Voucher	McKay Scholarships Program for Students with Disabilities
FL	Scholarship	Tax Credits for Scholarship Funding Organizations
GA	Voucher	Georgia Special Needs Scholarships
GA	Scholarship	Tax Credits for Student Scholarship Organizations
IA	Reimbursement	Tax Credits for Educational Expenses
IA	Donation	Tax Credits for School Tuition Organizations
IN	Scholarship	Special Needs Scholarship
IL	Reimbursement	Tax Credits for Educational Expenses
LA	Reimbursement	Personal Tax Deduction
LA	Voucher	New Orleans, Means Tested Voucher
LA	Voucher	Student Scholarships for Educational Excellence Program
ME	Voucher	Town Tuitioning Program
MN	Reimbursement	Tax Credits and Deductions for Educational Expenses
NC	Reimbursement	Tax Credit for Families with Special Need Student
OH	Voucher	Autism Scholarship Program
OH	Voucher	Cleveland Scholarship and Tutoring Program
OH	Voucher	Educational Choice Scholarship Program
OK	Scholarship	Tax Credit Scholarship Granting Organization
PA	Scholarship	Educational Improvement Tax Credit Program
RI	Scholarship	Corporate Tax Credits for Scholarship Organizations
UT	Voucher	Carson Smith Special Needs Scholarship Program
VT	Voucher	Town Tuitioning Program
WI	Voucher	Milwaukee Parental Choice Program

Source: The Friedman Foundation for Educational Choice. *School choice programs.* Retrieved from http://www.edchoice.org/School-Choice/School-Choice-Programs.aspx?id=12.

schools across the country (elementary: 21,870; secondary: 2,930; combined: 8,940). Of these private schools, 6,070 were Catholic schools (elementary: 4,640; secondary: 1,090; combined: 340).

Private school enrollment was 5,910,210 (elementary: 3,228,310; secondary: 827,390; combined: 1,854,510) compared with a public school enrollment of 49,386,000 (elementary: 34,285,564; secondary: 14,980,000). Of the nearly six million students in private schools, 2,119,341 were attending Catholic schools (elementary: 1,375,982; secondary: 593,097) (NCES, 2010).

In fiscal year 2007, the average per pupil expenditure in the U.S. public schools was $10,770, whereas private school tuition at that time averaged $6,600 (elementary: $5,049; secondary: $8,412; combined: $8,302). These figures have declined by several hundred dollars since the 2008 recession. However, Catholic school tuition during this period averaged $4,254 (elementary: $3,533; secondary: $6,046; combined: $5,801). Tuition is the single biggest source of funding for private schools, although grants, endowments and charitable giving from private sources are essential for most of these schools (NCES, 2010).

Private schools are overwhelmingly affiliated with a religious organization. The pattern of distribution of non-public schools among the states is not even. Several states in the Northeast serve almost 20 percent of their students in private schools, while some states in the West have less than 3 percent of their children enrolled in private schools. In their report for the National Center for Education Statistics, Broughman, Swaim and Keaton (2009) indicated the following:

> in the fall of 2007, there were 33,740 private elementary and secondary schools with 5,072,451 students and 456,266 full-time equivalent (FTE) teachers in the United States. The average private school size in 2007–08 was 150.3 students across all private schools. Private school size differed by instructional level. On average, elementary schools had 114.9 students, secondary schools had 282.0 students, and combined schools had 193.8 students. More private school students in 2007–08 were enrolled in schools located in cities (2,126,230), followed by those enrolled in suburban schools (1,987,714), followed by those in rural areas (607,095), and then by those in towns (350,602). Three-quarters (74.5 percent) of private school students in 2007–08 were White, non-Hispanic; 9.8 percent were Black, non-Hispanic; 9.6 percent were Hispanic, regardless of race; 5.4 percent were Asian/Pacific Islander; and .6 percent were American Indian/Alaska Native. Of the 306,605 private high school graduates in 2006–07, some 65.0 percent attended 4-year colleges by the fall of 2007. (p. 2)

State aid to private schools flows in a variety of ways and from many sources, but is not found in every state. Federal categorical funding, on the other hand, is available to private school students in all the states. However, not all private schools choose to apply for these funds. Two common methods of aid to private schools are through grants to the schools for supplemental services to students, and assistance to students directly in the form of non-instructional support.

For example, private schools with eligible populations may participate, on an equal basis, in federal categorical grant programs, like the federal Title I or Individuals with Disabilities Education Act (IDEA). State categorical programs allow for private school participation in some cases as well. Often such programs are coordinated through the local public school district or intermediate education agency. Several states provide transportation for children in private schools, supply textbooks for student use or subsidize subject matter instructional media that is free of religious content.

Support for private schools is justified from a variety of policy perspectives. Often, the first consideration is that the parents of private school children pay taxes for the support of the local public schools from which they do not benefit directly. In essence these families are doubly taxed, once to support the public schools and again to support the private school. Private schools are generally seen as a benefit to the community and as such are deserving of some public support. Also, each child in private school is one less child that requires full state support for education, thus relieving the state of that financial obligation. Imagine what would happen if every child currently enrolled in a private school showed up at his or her local public school at the beginning of the school term. Using the figures cited above for fiscal year 2007, that would mean 5,910,210 students, times an average of $10,770, for a total $63,239,247.

Charter Schools

The popularity of charter schools experienced a resurgence in the 1990s. Charter schools in America date back to pre-revolutionary times. In that earlier day, a charter would be granted to an individual or organization, often a church, to set up a school for the town's children. In this manner taxes would be collected by the town's elected officials, given to the chartering organization and used for support of the school (Kaestle, 1983).

Charter schools are publically funded, semi-autonomous entities, which operate under various formats (Nathan, 1999). Forty states "charter" such schools. Charter schools promote choice by offering parents a variety of educational settings and curricula from which to pick (Lubienski, 2003). Enrollments in charter schools reached 1,433,116 (NCES, 2010). Charter school legislation is a patchwork of policy from state to state with regards to which level of government is authorized to grant charters, the number of charter schools allowed to exist, who may seek a charter, how they are funded, how much revenue will be awarded the schools and the levels of autonomy and accountability under which the charter school operates. Table 7.2 displays this legal patchwork.

Carpenter (2006) conducted a study of 1,182 charter schools (87 percent of charters operating in 2001–02) and created a typology by categorizing the schools into five groupings based on their curricular approaches. Here is what Carpenter found:

1. *general*: includes "conversion" schools (29 percent of schools in sample);
2. *progressive*: schools that focus on individual student development approaches (29 percent);
3. *traditional*: schools emphasizing a "back-to-basics" approach (24 percent);
4. *vocational*: schools that equip students to transition from school to work (12 percent);
5. and *alternative delivery*: schools that provide most instruction outside a traditional bricks-and-mortar building, e.g., virtual schools (6 percent).

Dr. Carpenter found about three-quarters of charters do not target a specific student population for enrollment, while 26 percent serve students with specific needs or attributes.

Oversight of charter schools also varies by state, from fairly limited to close scrutiny. Who is authorized to grant and revoke charters is controlled by state statute. Therefore, one sees school district boards of education, the state education agency, a public university, a state charter school authority, municipalities and any number of variations and combinations awarding charters.

Table 7.2   Charter school laws across the United States.

State	Year law passed	Year law last amended	Number of charters operating (as of March 2008)	Number of charter schools allowed by state	State allows virtual charter schools[1]
United States	†	†	4,231	†	19[2]
Alabama	†	†	†	†	†
Alaska	1995	2001	25	60	Yes
Arizona	1994	2003	479	Unlimited	Yes
Arkansas	1995	2007	18	24[3]	Yes
California	1992	2007	703	100 new per year[4]	Yes
Colorado	1993	2007	140	Unlimited	Yes
Connecticut	1996	2006	19	24	No
Delaware	1995	2004	19	Unlimited	No
District of Columbia	1996	2005	74	20 per year	No
Florida	1996	2006	348	Unlimited	No
Georgia	1993	2007	65	Unlimited	No
Hawaii	1994	2007	29	48[5]	No
Idaho	1998	2005	30	6 new per year[6]	Yes
Illinois	1996	2005	61	60	Yes
Indiana	2001	2007	41	Unlimited[7]	Yes
Iowa	2002	2007	10	20	No
Kansas	1994	2004	30	Unlimited	Yes
Kentucky	†	†	†	†	†
Louisiana	1995	2004	54	42	No
Maine	†	†	†	†	†
Maryland	2003	No amendments	30	Unlimited[8]	No
Massachusetts	1993	2003	62	120[9]	No
Michigan	1995	2001	245	Unlimited[10]	No
Minnesota	1991	2006	148	Unlimited	Yes
Mississippi	1997	2005	1	15	No
Missouri	1998	2006	36	Unlimited[11]	No

Montana	†	†	†	†	†
Nebraska	†	†	†	†	†
Nevada	1997	2007	24	Unlimited[12]	Yes
New Hampshire	1995	2003	13	Unlimited[13]	Yes
New Jersey	1996	2002	56	Unlimited	No
New Mexico	1993	2006	66	75	Yes
New York	1998	2007	99	200[14]	No
North Carolina	1996	1998	103	100	No
North Dakota	†	†	†	†	†
Ohio	1997	2007	295	No new charters[15]	Yes
Oklahoma	1999	2007	15	3 new per year[16]	No
Oregon	1999	2005	81	Unlimited	Yes
Pennsylvania	1997	2002	132	Unlimited	Yes
Rhode Island	1995	2004	11	20	No
South Carolina	1996	2007	30	Unlimited	Yes
South Dakota	†	†	†	†	†
Tennessee	2002	2005	12	50	No
Texas	1995	2001	314	215[17]	No
Utah	1998	2007	60	Unlimited	No
Vermont	†	†	†	†	†
Virginia	1998	2004	3	Unlimited	No
Washington	†	†	†	†	†
West Virginia	†	†	†	†	†
Wisconsin	1993	2005	247	Unlimited	Yes
Wyoming	1995	2006	3	Unlimited	Yes

Source: Center for Education Reform (2008). Charter School Laws Across the States, Rankings and Scorecard: 2008. http://nces.ed.gov/programs/statereform/source.asp#cer.

† Not applicable.

[1] A virtual school, or cyber school, is a school that delivers academic instruction via the Internet or computer network to students in locations other than a classroom, supervised by a teacher who is physically present.

[2] The total reflects the number of "Yes" responses in the column.

[3] Arkansas allows 24 new start charter schools and unlimited conversion schools. Knowledge Is Power Program (KIPP) charter schools are exempt from the cap and may apply for licenses for additional open enrollment charter schools.

[4] California has an annual cap of 100 new charter schools that may open per year, and an absolute cap of 1,050.

[5] In Hawaii, one new school may be authorized for every new start that either has its charter revoked or has been accredited for three years or longer by an education accreditation authority.

[6] Although Idaho allows up to six new charters a year statewide, only one per school district is permitted each year, not including virtual charter schools. No whole school district may be converted to a charter district.

[7] Indiana allows for unlimited schools sponsored by local school boards, and 20 per year by the mayor of Indianapolis (increases by five annually).

[8] Maryland allows unlimited charter schools, but school districts create their own limits.

[9] Massachusetts must also approve three new charter schools in struggling districts.

[10] Michigan allows unlimited charters authorized by local school boards, intermediate school boards or community colleges. Although no single university may authorize more than 50 percent of the university total, charters that are authorized by state universities are limited to 150. Fifteen high schools in Detroit may be opened by groups meeting certain funding criteria.

[11] Missouri only allows unlimited charter schools in St. Louis and Kansas City.

[12] In Nevada, there is a moratorium on state-approved charters, and some districts also have moratoriums.

[13] New Hampshire allows unlimited charter schools authorized by local boards but limits state board-authorized schools to 20.

[14] New York allows for 200 new start charter schools, 50 of which are reserved for New York City and may be approved by any of the three authorizers (the State University of New York Board of Trustees, the Board of Regents and the chancellor), and unlimited conversion charter schools.

[15] Although Ohio does not allow new charters to open, charters meeting state performance standards are exempt and may open one new school for each school that meets the targets. Unlimited conversions may open; however, Ohio placed a moratorium on opening new virtual schools in 2005. Virtual schools operating prior to 2005 may remain open.

[16] As of January 1, 2008, only three new schools may be approved each year. Oklahoma, however, allows unlimited charter schools in districts with 5,000 or more students with a population of at least 500,000.

[17] In Texas, the total number of charter schools allowed excludes schools started by public universities.

From a school finance perspective, charter schools, in general, tend to be funded at a lower rate per pupil compared to "regular" public schools. In addition, issues related to school construction and ownership of real property are handled in different ways by each state. In some cases charter schools are on their own to raise money for school construction, while other states provide direct grants, and some states allow charter schools to participate in bond elections undertaken by the local public school district. Hiring, salaries and staffing decisions are left to the individual schools in most states. Similar to private schools, charter schools tend to pay teachers considerably less than public schools in their area. Disparities in salaries can often exceed as much as 30 percent.

Charter schools also tend to be much smaller than public schools, which raises the question of efficiency among some skeptics. Local school districts often view charter schools as a drain on resources as enrollment is siphoned off. This can be particularly devastating for rural school districts with small enrollments, where the loss of one hundred students to charter schools can be substantial.

A percentage of charter schools are run by for-profit corporate management organizations. Some advocates assert that profit-driven businesses will be more efficient at providing educational services and use tax dollars more wisely than school bureaucracies. Furthermore, the for-profit enterprises are touted as more nimble than school boards and can change more quickly to meet customer demands while better utilizing available resources (Walk, 2003).

Home Schooling

It is estimated that 1.5 to 2 million school-age children participated in home schooling in 2007 in the United States (National Household Education Surveys Program, 2007). Home schooling is the oldest form of childhood education, which existed, of course, even before there were schools. All states have policies that allow some form of home education, which range from very liberal "hands–off" policies to policies that require thorough and detailed educational plans or supervision by a third party, for example, a private school or state-licensed teacher.

Some states encourage home-schooling families to access courses and programs available in their local public schools (Bauman, 2002). After all, these families pay taxes to support the public schools. Iowa, for example, allows schools districts to count home-schoolers as part-time students for enrollment count purposes if the school district provides services to the families or allows the children to participate in selected classes. Music, art, foreign language, extracurricular teams and activities are examples of areas where home-schooled children engage with the local public schools. School districts may even provide assistance with curriculum development, textbooks and library/media services. As with other forms of choice, the million-plus students in home education do not draw full state support for their education, thus reducing the total state obligation for pre-collegiate education.

Summary

In this chapter a variety of alternative systems to the traditional public schools have been explored. Manifestations of choice policy in the form of vouchers, charter schools, direct and indirect government aid to private schools and home schooling are examples of alternative systems covered in the chapter. Choice policy based on the child benefit theory was considered in light of these education policies and their effect on education finance. Some choice policies promote reduced government expenditures, while others create inefficiencies. School choice represents a relatively new policy area, as many of the current school choice policies are less than

Table 7.3   Percentage of home-schooled students, ages five through seventeen, with a grade equivalent of kindergarten through twelfth grade, by school enrollment status: 1999, 2003 and 2007.

School enrollment status	Year					
	1999		2003		2007	
	Percent	+/−	Percent	+/−	Percent	+/−
Total	100	†	100	†	100	†
Homeschooled only	86	6	82	7	84	5
Enrolled in school part time	18	6	18	7	16	5
Enrolled in school for less than 9 hours a week	13	6	12	6	44	5
Enrolled in school for 9 to 25 hours a week	5	3	6	4!	5	3!

† Not applicable.
+/- is margin of error for a 95 percent confidence interval.
! The standard error for this estimate is greater than 30 percent of the estimate. Interpret with caution.
Note: Excludes students who were enrolled in public or private school for more than twenty-five hours a week and students who were homeschooled primarily because of a temporary illness. Detail may not sum to totals because of rounding.
Source: U.S. Department of Education, National Center for Education Statistics, Parent Survey of the 1999 National Household Education Surveys Program (NHES); Parent and Family Involvement in Education Survey of the 2003 and 2007 NHES.

two decades old. Nor have they been fully unleashed. The full financial impact of these policies has yet to be determined.

References

Bauman, K. J. (May 2002). Home schooling in the United States: Trends and characteristics. *Education Policy Analysis Archives*, *10*(26). Retrieved from http://epaa.asu.edu/epaa/v10n26.html.

Broughman, S. P., Swaim, N. L., and Keaton, P. W. (2009). *Characteristics of private schools in the United States: Results from the 2007–08 private school universe survey* (NCES 2009-313). Washington, DC: National Center for Education Statistics, Institute of Education Sciences, U.S. Department of Education.

Carpenter, D. M. (2006). Modeling the charter school landscape. *Journal of School Choice*, *1*(2), 47–82.

Friedman, M., and Friedman, R. (1980). *Free to choose: A personal statement.* New York: Harcourt Brace Jovanovich.

Kaestle, C. F. (1983). *Pillars of the republic: Common schools and American society, 1780–1860.* New York: Hill and Wang.

Lubienski, C. (2003). Innovation in education markets: Theory and evidence on the impact of competition and choice in charter schools. *American Educational Research Journal*, *40*(2), 395–443.

Nathan, J. (1999). *Charter schools: Creating hope and opportunity for American education.* San Francisco: Jossey-Bass Publishers.

National Center for Education Statistics (2007). *Digest of education statistics.* Washington, DC: National Center for Education Statistics.

National Household Education Surveys Program (2007). *Parent and family involvement in education.* Washington, DC: National Center for Education Statistics. Retrieved from http://nces.ed.gov/pubs2009/2009030.pdf

Pierce v. Society of Sisters, 268 U.S. 510. (1925).

Walk, D. R., Jr. (2003). How educational management companies serve charter schools and their students. *Journal of Law and Education, 32*(2), 241–42.

Wisconsin v. Yoder, 406 U.S. 205. (1972).

Zelman V. Simmons-Harris, 536 U.S. 639. (2002).

Federal Funding for Education

8

Aim of the Chapter

IN THIS CHAPTER THE READER WILL GAIN A SENSE of the size and scope of federal education initiatives. Particular attention is paid to several major pre-collegiate (PK–12) and postsecondary education programs, their funding levels and policy goals. A sampling of education programs from federal agencies other than the U.S. Department of Education is also presented.

Introduction

When the founders of the nation crafted the U.S. Constitution, they did not devote an article or section to the question of education. While education was seen by many of the founders as critical to the viability of the new democratic republic, the idea of a national system of education was not envisioned within the federal role. At that time, the late eighteenth and early nineteenth centuries, education was seen as a state function by some, a church function by others and a private family matter by most. Despite these views, efforts by the national government to encourage states to assume responsibility for a system of education can be seen as early as the 1780s, well before the U.S. Constitution was adopted, in such legislation as the Northwest land ordinances (see chapter 2). Several of the thirteen original states adopted education articles in their initial constitutions, thus taking up the role of education as a state function.

Despite this federal system of divided roles and responsibilities between the states and the national government, education has always had some federal involvement, often in ways not thought about much today. For example, West Point, the U.S. Military Academy, was created in 1802 by the national government and signed into law by Thomas Jefferson. Treaties negotiated between the United States and indigenous tribes or "nations" sometimes had an education clause, offered as an inducement or requested by the native peoples as a condition for signing. Today,

the federal government devotes barely 3 percent of its $3 trillion budget to educa-
tion, although this adds up to almost $260 billion. Federal contributions to PK–12
education account for about 9 percent of the half-trillion dollars spent annually
by all levels of government. Funding for higher education is also significant, yet
small relative to state support and student tuition (U.S. Department of Education,
2011). No fewer than sixteen federal departments have some education funding
and program responsibilities (NCES, 2010). The footnote at the end of the chapter
leads to a comprehensive list of federal programs with an educational mission. For a
country without a national system of education, the U.S. federal government runs
a vast education enterprise that is literally global in scale and reaches into every
classroom in America.

Programs within the U.S. Department of Education

The U.S. Department of Education of today was established in 1980, and with it
the cabinet position of secretary of education. From 1953–79 federal education was
headquartered within the Department of Health, Education, and Welfare in the
Office of Education, which had a commissioner of education. Although numerous federal education initiatives had been undertaken over the years—for example,
the Morrill Act of 1862 for land grant colleges and the Freedmen's Bureau in
1865 within the War Department—it wasn't until 1867 that the U.S. Bureau of
Education and commissioner of education were established in the Department of
Interior. The bureau's role was to collect education statistics and oversee the im-
mense education land grant programs within the states and territories. Today the
U.S. Department of Education is the key federal agency concerned with national
education policy across the P–20 education endeavor in the United States (U.S.
Department of Education, 2008). Below is a listing of some of the more notable
programs administered through the U.S. Department of Education.

The Elementary and Secondary Education Act (ESEA)

The ESEA law was enacted in 1965 as part of a grand federal policy design known
as the "Great Society," which was President Lyndon B. Johnson's (1963–69) vision for a new America. The ESEA effort linked education, civil rights legislation and poverty reduction programs into a policy framework that spread across
the nation. *Brown v. Board of Education of Topeka, Kansas* (1954) declared an end
to segregated schools in America, but integration was slow in coming. Previous
presidents Dwight D. Eisenhower (1953–61) and John F. Kennedy (1961–63)
had to use federal troops on occasion to enforce court orders for school integration. President Johnson saw the ESEA as accomplishing many things, including
serving as a lever to move recalcitrant states and school districts away from their
segregated dual systems of schools. The new money flowing to schools was wel-
comed and needed.

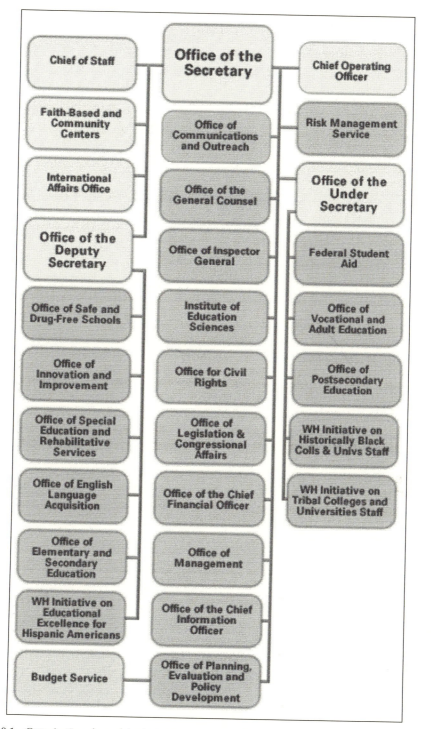

Figure 8.1   Organization chart of the U.S. Department of Education.
Retrieved from http://www.ed.gov/print/about/offices/or/index.html.

With the passage of the new law, massive infusions of new funding, relative to previous federal appropriations, headed toward schools. Soon the threat of a loss of federal funding facilitated the process of having school districts comply with an avalanche of federal regulations. The aim of the program then, as it is today, was to mitigate the adverse educational effects of poverty for children from disadvantaged backgrounds. This is particularly the case for Title I of the act. President Johnson's vision saw education as a vehicle for lifting up the poor and a means of enabling the previously disenfranchised to enter mainstream American life.

In the long history of the program there have been many ups and downs, as some presidents and some members of Congress have tried to reduce its impact or eliminate the program altogether. Despite these efforts the basic focus of the program has remained the same: improving the educational chances of children from low-income families. Not every president since Johnson, Richard M. Nixon (1969–74) and Ronald Reagan (1981–89), for example, has been enthusiastic about the ESEA, whether because of the cost of the program or the belief that the ESEA is an excessive federal intrusion into a state matter, i.e., education. But efforts to scuttle this huge federal education program have all failed. One reason for this is that almost every school district in the nation and every congressional district receives some ESEA funding. Thus, the constituency for the program is ubiquitous.

President George H. W. Bush (1989–93) and Congress started the process of expanding the scope of the policy influence of the ESEA when he grafted his America 2000 initiative on to the law. This began the federal push for educational standards and accountability systems in each of the states for Title I students. The policy push was an effort to show results for the money being spent. President William J. Clinton (1993–2001) continued the process of using the ESEA as a leverage point to broaden the influence of the federal government in the educational matters of the states. His Goals 2000 program was added to the ESEA, expanding the standards and accountability system to a statewide system affecting all schools and students, regardless of whether they were ESEA Title I students. In this iteration of the law, Title I students and non-Title I students would be tested in selected grades to measure whether poor children were getting the same quality education as children from more affluent families.

President George W. Bush (2001–09) ushered in the 2001 reauthorization of the Elementary and Secondary Education Act and renamed it the No Child Left Behind Act (NCLB). This education policy piece solidified the ground gained in the two previous reauthorizations and expanded federal influence into mandatory prescriptive annual testing programs for almost all students, numerical goals for increased student achievement for all students, teacher qualifications requirements and school choice options for parents.

The appropriation for fiscal year 2010 for NCLB was almost $20 billion. The NCLB, or ESEA, program is divided into several subunits, the most well-known of which is Title I. Here is a brief description of the major components of Title I.

TITLE I–PART 1-A   Title I–Part 1-A distributed approximately $15.5 billion in fiscal year 2010 (FY10) to school districts through a multitiered formula that uses a weighted factor per child depending on the relative concentration of poor children in the school district. The Basic Grant will allocate $6.8 billion to all qualifying school districts based on the number of poor children in the school district. Census data are used to make these calculations. Another $1.4 billion is given to school districts with 15 percent, or 6,500 children, living in poverty through Concentration Grants. Targeted Grants are issued to school districts with poverty thresholds above 16 percent, and are enriched further at the 38 percent level. Districts with more than 35,500 low-income students get about three times the basic formula per child (New America Foundation, 2008).

TITLE I–PART 1-C   Title I–Part 1-C is a program devoted to the children of migratory agricultural workers. This section of Title I awarded school districts $394 million in FY10 to help mitigate the effects of families who migrate to earn a living as agricultural workers. Eligible students included children whose parents work on farms or in fishing, forestry and agriculturally related industries.

TITLE I–PART 1-D   Title I–Part 1-D serves delinquent children and youth who have been adjudicated by the state and are incarcerated. Neglected children, who are wards of the state, also receive some benefit through Part D. In FY10 $50 million went to Part D funding.

TITLE VIII–IMPACT AID PROGRAM   The Title VIII–Impact Aid Program has existed since 1950 and was established by Congress to mitigate the economic effects

Table 8.1   Title I student weighted formula elements.

School district % of poverty	Per student weight
0–15.6%	1
15.6–22.1	1.75
22.1–30.2	2.25
30.2–38.2	3.25
>38.2	4

School district number of poverty	Per student weight
1–691	1
692–2,262	1.5
2,263–7,851	2
7,852–35,514	2.5
>35,515	3

Source: New America Foundation, 2012.

of federal activity within school districts. The program was folded into the ESEA as Title VIII. In consideration of the fact that the federal government pays no state or local taxes, Impact Aid is granted to eligible school districts because of the existence of nontaxable federal land, personnel who live or work on federal property, and stores on federal property that do not generate local or state sales tax. Military bases, Indian reservations and national forests are examples of such property.

School districts are awarded funds based on several criteria: nontaxable federal land; basic support payments for "federally connected" students whose parents or guardians either live or work on federal property; heavily impacted school districts, which have high levels of federally connected children and low levels of local support; Indian land; and military connected students who are disabled. The program also has a limited school maintenance and construction support program. In FY10 about $1.25 billion was set aside for Impact Aid.

TITLE II–SCHOOL IMPROVEMENT Title II–School Improvement is also known as Title II, a subsection of the ESEA. It covers an array of grant programs from teacher quality, to gifted education, to rural education. Many of the programs included in this section of the ESEA existed in other legislation and were included in the ESEA bill as a result of consolidation efforts. The appropriation for School Improvement in FY10 was $5.2 billion.

TITLE III–BILINGUAL EDUCATION Title III–Bilingual Education is centered in the Office of English Language Acquisition and a part of the ESEA as Title III. This program provides grants to school districts. For FY10 the appropriation was $700 million. About 4 million students nationwide are new learners of English and need English acquisition learning assistance.

Several other smaller programs, such as Reading First, round out the ESEA programs.

Individuals with Disabilities Education Act (IDEA)

The origin of federal policies and programs for students with disabilities has an interesting background filled with many surprising circumstances. For example, President Abraham Lincoln in 1864 signed a bill that helped today's Gallaudet University become a degree-granting institution (Gallaudet University, 2008). But the real changes to federal policies and programs for the disabled came out of the legal struggles for civil and equal rights during the 1960s and 1970s. Plaintiffs in cases like *Pennsylvania Association for Retarded Children (PARC) v. Commonwealth of Pennsylvania* (1972) and *Mills v. Board of Education of the District of Columbia* (1972) affirmed the rights of children with handicapping conditions to access a free education in the public schools. By 1973 Congress passed the Vocational Rehabilitation Act, which barred discrimination against the disabled; in essence, it established constitutionally protected civil rights for the disabled. Section 504 of the act specifically applied to organizations receiving federal funds, such as schools and colleges. But Section 504 of the Vocational Rehabilitation Act had no money attached to it.

Congress passed Public Law 94-142, the Education for All Handicapped Children Act, in 1975, and since then services to children with disabilities have expanded immensely. The law, along with Section 504, threw open the schoolhouse doors to students who had previously been excluded or marginalized. Today's version of the law is known as IDEA. In what has proved to be a prescient declaration, President Gerald Ford (1974–77) foretold the problems the nation would face with this watershed law. Here is his signing statement:

> I have approved S. 6, the Education for All Handicapped Children Act of 1975.
>
> Unfortunately, this bill promises more than the Federal Government can deliver, and its good intentions could be thwarted by the many unwise provisions it contains. Everyone can agree with the objective stated in the title of this bill—educating all handicapped children in our Nation. The key question is whether the bill will really accomplish that objective.
>
> Even the strongest supporters of this measure know as well as I that they are falsely raising the expectations of the groups affected by claiming authorization levels which are excessive and unrealistic.
>
> Despite my strong support for full educational opportunities for our handicapped children, the funding levels proposed in this bill will simply not be possible if Federal expenditures are to be brought under control and a balanced budget achieved over the next few years.
>
> There are other features in the bill which I believe to be objectionable and which should be changed. It contains a vast array of detailed, complex, and costly administrative requirements which would unnecessarily assert Federal control over traditional State and local government functions. It establishes complex requirements under which tax dollars would be used to support administrative paperwork and not educational programs. Unfortunately, these requirements will remain in effect even though the Congress appropriates far less than the amounts contemplated in S. 6.
>
> Fortunately, since the provisions of this bill will not become fully effective until fiscal year 1978, there is time to revise the legislation and come up with a program that is effective and realistic. I will work with the Congress to use this time to design a program which will recognize the proper Federal role in helping States and localities fulfill their responsibilities in educating handicapped children. The Administration will send amendments to the Congress that will accomplish this purpose. (Ford, 1975)

The controversy over adequate federal funding, excessive paperwork and federal intrusion into state and local matters continues to this day. The original law called for funding authorization levels equal to 40 percent of the excess cost of educating a child with disabilities. Debates over establishing a figure for excess cost have never ended. During the three-plus decades of the law's existence, federal appropriations for Public Law 94-142, today known as IDEA, have never reached above 18.5 percent of the calculated excess cost, the figure in FY05. The percentage of excess cost appropriated by the Congress decreased in FY06, FY07 and FY08. In FY10, Congress appropriated just over $12.5 billion under the act. Below is a breakdown of the more significant pieces of the law.

IDEA–PART B   This portion of the law provides grants to states and school districts through a formula that awards dollars per child with an individualized education program (IEP). Per pupil amounts will vary from year to year. Each recipient must assure the federal government that it will comply with the provisions of Section 504 and IDEA, which include:

- A free appropriate education for all disabled children. Appropriate has been interpreted to mean "to established public standards."
- A proper assessment of the educational needs of the child.
- The provision of related services needed to benefit from the planned education.
- Informed consent and parental involvement at each stage of assessment and education placement.
- Covers preschool through high school.
- Offered according to an individualized education program.

Numerous rights and due process procedures are also afforded disabled students and their parents under the law. Almost the entire $12.5 billion appropriation goes to Part B, grants to states. These funds are in turn distributed to school districts. Part D—National Activities to Improve Education of Children with Disabilities provides smaller grants, some on a competitive basis, for everything from personnel development to parent information activities. Vocational Rehabilitation Services adds another $3.2 billion to this special needs population.

The federal government has never come close to fulfilling its promise to fund 40 percent of the excess cost; nevertheless, states and school districts have no choice but to provide the free education outlined in the law, regardless of the expense. As a nation America adopted a policy to expand educational opportunity to all children with disabilities—and rightly so. But the national government has failed to provide sufficient funding to states and school districts to meet the challenge set before them.

Vocational Education

This federal program is now called career and technical education and has roots that trace back to the early twentieth century (the Smith Hughes Act of 1917, for example). Appropriations in FY10 reached almost $2 billion. State grants for career and technical programs as well as adult basic education are funded through this initiative.

Higher Education Student Financial Aid   In FY10 the federal government appropriated slightly less than $50 billion for grants and loans to higher education students and their families. Of this amount about $23 billion was distributed in the form of grants to students. Grants were given to students mostly on a need basis, although some of these funds were awarded on merit. Grants, unlike loans, are not repaid. A more complicated picture emerges when looking at the higher education loan part of the budget. Direct loans from the government have declined over the past half-dozen years because of how the program has been structured and incentives to banks to get involved in student lending. Loan guarantees, tax breaks and work-study programs are also part of the financial aid picture, but the federal budget only reflects part of this total spending.

Higher Education Act (HEA)

As the name implies the HEA is concerned with postsecondary education. Much smaller than the ESEA, the HEA spent $2.6 billion in FY10. Programs in the HEA fall under two broad categories: "strengthening institutions" and "improving post-secondary education," which includes a variety of earmarks for minority-serving institutions. About $1 billion is set aside to facilitate transition to college for first-generation college students and another half-billion for a list of scholarship programs, for example, Senator Robert C. Byrd, Senator Jacob Javits and Thurgood Marshall scholarships.

Historically Black Colleges and Gallaudet University

The federal government has unique relationships with several institutions of higher education. Howard University, founded during the Reconstruction Era under the Freedmen's Bureau, and Gallaudet University, also established in the nineteenth century, are two examples. During FY10 about $500 million was spent on such schools and programs. Combined, the Office of Postsecondary Education distributed about $2.6 billion in FY10, which was added to a similar amount under the HEA. In addition, a substantial college loan and grant program is funded by the federal government through the U.S. Department of Education, which reaches upwards of $46 billion.

Summary for U.S. Ed.

Total appropriation to the U.S. Department of Education for FY10 reached almost $77 billion (U.S. Department of Education, 2011). Yet despite this substantial sum, only about half the money spent by the federal government on education-related programs comes from the U.S. Department of Education. An overview of some of the more notable education-related programs not in that department is covered in the next part of this chapter.

Federal Programs Not Under U.S. Ed.

A large number of education or education-related programs run by the federal government are not found within the U.S. Department of Education. In fact, of the approximately $170 billion spent by the federal government on education-related matters, about $100 billion is appropriated to agencies other than the U.S. Department of Education. Below is an abbreviated list of some of these programs. The list is presented in an effort to show the scope and variety of federal programs. Figures used relative to program spending are for Fiscal Years 2008 through 2011. A more complete list of programs and expenditures is displayed online at the NCES Website.[1]

United States Department of Agriculture—$15.5 Billion

The origins of the school nutrition programs trace back to the Great Depression. At that time millions of children were going hungry while farm products went to waste for lack of a market. The government intervened to buy up the surplus commodities and serve them as school lunches to children. In the process, thousands of people, mostly women, were put to work serving meals in the new school lunch program. Within a few short years, millions of students and tens of thousands of schools were participating in the program. In addition to the National School Lunch Program, there is the School Breakfast Program, Child and Adult Care Food Program, Summer Food Service Program, Fresh Fruit and Vegetable Program, and afterschool snack components to the program (United States Department of Agriculture, 2012).

Department of Health and Human Services, Head Start Program—$7 Billion

This preschool program is yet another legacy of President Johnson's Great Society. The program has provided early education services to poor and low-income families since the mid-1960s. States receive grants from the federal government to provide child and family educational support. Common practice is to have Head Start programs administered through state and local human services agencies.

Department of the Interior, Bureau of Indian Affairs, Bureau of Indian Education—$850 Million

The relationship between the native peoples of America and the United States government is a long and complex one. Many school-age children whose background is from the original American people do not live in a school district, or for that matter, within the jurisdiction of a state. The lands that have been set aside through negotiations and treaties for the many native populations are considered federal lands, even though they are within the boundaries of a state. Many such groups have their own independent schools, and many Native American children attend regular public school. However, the Office of Indian Education oversees 183 PK–12 schools and 26 tribal postsecondary institutions, and serves 238 distinct tribes in 23 states and 64 reservations. PK–12 spending exceeded $700 million, while higher education allocations were over $100 million (Bureau of Indian Education, 2011).

Department of Defense, Service Academies, Tuition Assistance for Service Personnel, Junior and Senior ROTC, Dependent Schools—$6.7 Billion

Appropriations for defense-related education expenses have been made since the nation was founded. Today the federal government allocates dollars in support of the three service academies: Army, Navy and Air Force. The Coast Guard Academy is funded through the Department of Homeland Security.

The Department of Defense Dependent Schools, a PK–12 system, spend over $1.8 billion to run 194 schools or pay tuition for dependent children in the United States and around the globe. The program pays 8,700 educators to serve over 86,000 children. It even includes one school in Cuba. Domestically, $300 million is spent running Department of Defense schools (Department of Defense Education Activity, 2011).

The need for leaders within the military goes beyond the service academies. In fact, almost three times as much is spent on the Reserve Officers' Training Corps to pay for the Junior Corps and Senior Corps. Additionally, funding is spent on tuition assistance and professional development for military personnel for everything from flight training to law school.

Department of Veterans Affairs–$4.6 Billion

This department exists separately from the Department of Defense and was created to serve the needs of former military personnel, i.e., veterans. Among the more notable programs under this agency are: Vocational Rehabilitation for Disabled Veterans; All-Volunteer Force Educational Assistance; Veterans Educational Assistance and Non-College Job Training; Reserve Educational Assistance; and Dependents' Educational Assistance.

Summary

It is important to note that despite the fact that the United States does not have a national education system, the uniformity of the PK–12 education systems from state to state is quite remarkable. Some will argue this point on parochial grounds, but the similarities are hard to dispute among programs that are supposed to be within the domain of the individual states. Consider these many examples. On average, most states require around 180 days of instruction. Overwhelmingly, the most common grade organization is elementary school, middle school and high school. Professional licensing, though varying somewhat from state to state, is everywhere. Curricular offerings and sequencing are remarkably similar across the country. Even the textbooks and instructional materials are common to many states. Thus, one hears little complaint from our highly mobile society about the need to make the system more uniform across the country. Families moving to a new state or within a state understand that there will be some new or different aspects to their new school. However, they don't expect, and rarely get, a radical change from what they knew in their former community.

In most instances this uniformity has evolved because of the iterative process of educational change and innovation moving from school to school, school district to school district and state to state. Common practice among political leaders sees education policies developed in one state and moved to and adopted by other states. This practice of copying one's neighbor has gone on since the beginning of the nation. Legal thinking and court decisions from all levels have contributed to shaping the system of schooling in the nation as well. Colleges of education and professional organizations also add to the development and structure of the pre-collegiate system. Most of these changes have been slow, subtle and voluntary. They have spread in a somewhat organic fashion, many without government mandates.

The influence of the federal government in designing the PK–12 education system has historically been restrained and removed from direct influence. The land ordinances of the 1780s, for example, though involving millions of acres of land and vast sections of the nation, were accomplished with a piece of legislation comprised of a few sentences. Today, states and school districts grapple with laws like the No Child Left Behind Act of 2001, which comprised 1,200 pages of mandates and procedures that states and school districts must obey.

If the federal government were a school district, it would rank among the largest in the nation. If the federal government were a system of higher education, it would rival that of all but the most populous states. The size and scope of direct and indirect education initiatives on the part of the United States government is vast. Many of the most significant advances in American education can be traced back to some stimulus from federal education policy.

Clearly, there is a role for the federal government in education; few dispute this assertion. But the nature and extent of this role is difficult to define. Should the federal government be limited to educational matters that involve the U.S. Constitution, for example, student rights issues? Does the federal role also involve promoting the "national interest?" If so, how is the national interest defined and who defines it?

Federal financial assistance to pre-collegiate education accounts for about 9 percent of total spending and varies from a low of 4 percent in New Jersey to 18 percent in Alaska. The loss, or even decline, of federal funds for any state would be significant. However, given the extent and nature of federal involvement in PK–12 education, more policy leaders are asserting that perhaps it is time for a national dialogue about the role of the national government. What do you think?

Note

1. http://nces.ed.gov/programs/digest/d09/tables/dt09_375.asp.

References

Brown v. Board of Education of Topeka, Kansas, 347 U.S. 483, 74 S. Ct. 686. (1954).

Bureau of Indian Education (2011). U.S. Department of the Interior. Retrieved from http://www.bie.edu/.

Department of Defense Education Activity (2011). Washington, DC: U.S. Department of Defense. Retrieved from http://www.dodea.edu/home/facts.cfm?cId=budget.U.S.

Ford, G. (1975). *President Gerald R. Ford's statement on signing the Education for All Handicapped Children Act of 1975*. Retrieved from http://www.ford.utexas.edu/LIBRARY/speeches/750707.htm.

Mills v. Board of Education of the District of Columbia, 348 F. Supp. 866 (D.D.C., 1972).

National Center for Education Statistics (2010). *Digest of education statistics*. Washington, DC: National Center for Education Statistics.

New America Foundation (2012). Retrieved from http://www.newamerica.net/background-analysis/no-child-left-behind-act-title-i-distribution.

Pennsylvania Association for Retarded Children (PARC) v. Commonwealth of Pennsylvania, 343 F. Supp. 279 (E.D. Pa., 1972).

U.S. Department of Agriculture (2012). *National School Lunch Program*. Washington, DC: U.S. Department of Agriculture. Retrieved from www.fns.usda.gov/cnd/Lunch.

U.S. Department of Education (2008). *Federal role in education*. Retrieved from http://www.ed.gov/print/about/overview/fed/role.html.

U.S. Department of Education (2011). *Department of Education budget*. Retrieved from http://www2.ed.gov/about/overview/budget/budget12/12actioncr.pdf.

Employee Compensation

9

Aim of the Chapter

THIS CHAPTER EXPLORES THE COMPELLING REASONS for having compensation policies and systems that successfully attract and retain high-quality personnel to public education. Special attention is paid to the central role of teachers and how American society has grappled with the relationship between a desire for effective pre-collegiate education and a willingness to compensate those interested in teaching. The scope and nature of employee compensation systems are examined along with the emerging policy alternatives.

Introduction

Education is a labor-intensive enterprise, and as a result, employee salary and benefits comprise the largest expense in a school or school district's operating budget. The struggle to pay school employees traces back to the earliest days of schools in America, when many towns and villages literally had to pass the hat to scrape up the cash to pay for a teacher. During those early days, cash was in short supply, so it was common to pay teachers with room and board and little else (Kaestle, 1983). The struggle continues to this day.

The ability to build and maintain a compensation structure that successfully attracts and retains qualified personnel is essential to school success. As the system of education became more complex through the years, the number and types of personnel required to offer a quality education also expanded. Each generation has added to this expansion. From the itinerant teacher traveling the countryside with his books, who was hired by a group of families for a few months to teach children basic literacy and numeracy, to the staffing of the first grammar schools in cities and towns, to the many specialists and technical employees found in school districts today, raising money for salary and benefits in order to hire needed staff remains a key organization function.

Today we see one employee on the payroll for every nine students enrolled in the school district (NCES, 2011). Back in time, school meant a teacher and a group of children. As the schools grew larger, a principal was added. As the town added more schools, the superintendent position was created. Communities decided that transportation was a service that schools should provide, and drivers and mechanics were added to the payroll. School lunch, nursing, libraries, guidance and counseling, and a host of other programs and services were added to the functions and duties of the schools. Newly added positions found in school districts include computer technicians, accountants, lawyers and news media specialists.

Of course, teachers remain the heart and soul of the education process and the most important personnel in the educational success of students. So despite all the changes to schools, and all the additional personnel, drawing the best available people to teaching is an ongoing endeavor. The development of the teaching profession in America, coupled with the ever-shifting societal landscape, has only served to make the endeavor a more complex struggle.

The Scope of the Challenge

It is appropriate to start this chapter by looking at teacher compensation. As a group, teachers represent the largest collection of employees in a school district—a little over half. They are also allocated the largest share of salary and benefits. But teacher pay has been and remains a problematic area for policy leaders, administrators and teachers alike. Specifically, teacher salaries are controversial because of historical, societal and economic factors that affect how teachers are compensated.

Issues of teacher compensation have run concurrently with the development of the public school system in the United States. The nineteenth-century movement for universal free public education had many obstacles to overcome. Among these challenges were funding for operational expenses, funding for facilities and the question of who would staff the schools and teach the millions of youngsters eager to enroll. Reformers and societal leaders supported the idea that young women could serve in that capacity.

Many of the advocates for universal free education had a ready response for the critics of their quest when it was pointed out that communities could never afford the salaries for all the teachers needed for all the new schools. The reformers countered that the solution lay in hiring teenage girls who had recently completed their education. These girls—and they were girls—could be hired at a salary much lower than the teachers of that day, often men. The idea was that teaching could be a nice avocation for the girls in the interim between completing their studies and getting married (Kaestle, 1983).

But other societal leaders had a different view of teaching. The women's movement of that same era redefined the role of women in America, as forces like the western frontier, urbanization and the Civil War changed the country. Leaders like Catharine Beecher advanced the idea that teaching was a suitable and honorable profession for a single young woman to pursue. She encouraged young ladies to heed the call and venture west to staff the schools springing up in the new towns and cities. Such service would be a benefit to the individual and a service to the nation (Webb, Metha and Jordan, 2007).

Thus, teaching in the United States developed as a "women's profession," and the legacy of those origins is felt to this day. While sexism remains a point of contention in American society today, particularly over matters of equal pay for equal work (CNN, 2009), the outrages of the nineteenth century are often hard to understand from the modern perspective. Women were excluded from most professions. This was achieved by simply not admitting them to professional schools or barring them from professional licenses. Pay differentials were routinely and customarily assigned based on gender. Men were paid more than women for doing the same job.

Women in teaching were mostly viewed as temporary workers who would leave their position, or be forced out by the school board or superintendent, upon marriage or starting a family. The stereotypical spinster schoolmarm was the exception. Such restrictions eventually faded away as the need for teachers grew and the education system expanded to the elementary and secondary schools we know today.

The legacy of these earlier times shows up in the compensation of teachers today. In their study of teacher salaries, *The Teaching Penalty: Teacher Pay Losing Ground*, Allegretto, Corcoran and Mishel (2008) describe how teaching as a profession is undercompensated compared to professions that require similar kinds of preparation like accounting, social work and government office workers. The study does a good job of exploding the myths about teaching as it looks at salary comparisons based on yearly, monthly, weekly and even hourly measures. The argument that teachers don't work a full year or a full eight-hour day is clearly dispelled. They even examine the charge that teachers get much better benefits than most other professionals and demonstrate how this does not help achieve parity.

It was projected that 3.7 million people were employed as teachers in 2010, and public schools accounted for 3.2 million. Fifty-one percent of the staff in schools are teachers, and 71 percent of teachers are women. The National Center for Education Statistics (2011) reports that the average teacher salary across the nation for the school year 2008–09 was $53,910. In an earlier study, the American Federation of Teachers (AFT, 2008) reported an average salary of $51,009 for the 2006–07 academic year, yet they also state that teachers earn 70 percent of what other professionals with similar qualifications earn. The report notes a variation of over $6,000 in this average between states with the lowest to the highest average. The AFT report also surveyed charter school teachers in twenty-nine of the forty states with charter school laws and reported an average salary of $41,106 for those teachers.

Today, the consequences of this salary lag are seen in teacher shortages in some geographic areas, and shortages in teachers of mathematics, science, special education and language minority students. Part of the phenomena of teacher shortages is explained by another societal shift that started in the post-World War II era and accelerated during the civil rights movement of the 1960s and 1970s. As barriers to discrimination against women in the workplace and the professions started to fall, more women opted to seek careers in fields other than teaching.

The economic recession of 2008–10 eased the teacher shortage substantially. But if past history is a guide, school districts can expect a return to teacher shortages as the economy rebounds. Here again, the value of having a long-range view about personnel matters is critical to successfully staffing schools and school districts. Economic conditions are always changing, and effective school leaders and policy makers understand this and plan accordingly.

A related problem is also seen today in the number of teachers who leave teaching early in their careers for other professional pursuits. While pre-collegiate teaching remains an overwhelmingly female-dominated field by about three to one, the seemingly endless pool of capable young women available for teaching jobs in the late nineteenth and early twentieth centuries has dried up. Today, schools find themselves competing for talent along with every other profession.

Some researchers (Hoxby and Leigh, 2005) have even pointed out that another consequence of expanded career opportunities for women has been the decline of entrants to teaching from selective colleges and universities. As teaching developed into a profession on par with other professions and the system matured, qualifications to be admitted to teaching also increased. Such requirements moved from an eighth-grade education to teach elementary school, to a high school diploma, to a year or two at a teacher's college or normal school, to a bachelor's degree. In the past, many teachers, particularly women, graduated from colleges and universities that had highly competitive entrance requirements. Today, more and more teachers come from schools with less selective criteria. The implication is that a brain drain of sorts from the teaching profession has resulted.

The decline in the number of people pursuing teaching as a career and the number of undergraduates preparing to become teachers has diminished the pool of choices available to school districts, charter schools and private schools looking for teachers. Historically, the solution to such a problem has been to lower or modify entrance requirements to the profession. States establish "emergency" credentials for people willing to teach and who promise to get fully credentialed at some later date. The No Child Left Behind Act (NCLB) of 2001 was supposed to eliminate this problem. But some schools and school districts must still staff classrooms with people who are teaching out of their field or who are totally unqualified for the job. When the children show up, someone has to take over the class.

Alternative credentialing programs are another common means of meeting teacher shortages. Such programs typically offer paths to the teaching profession for those who may already have a bachelor's degree and did not complete a teacher education program, but wish to transition into teaching. Some school districts with particularly difficult recruiting challenges have resorted to "grow your own" programs, wherein district employees, like a classroom aide or clerical person, are assisted in attaining a bachelor's degree and teacher licensing. The nature and quality of such programs vary greatly. A contemporary alternative licensure program is Troops to Teachers, for example. As the name implies, it is designed to help service personnel transition to the education profession. The program is popular in many school districts and has proven successful in getting mature and highly skilled people into the classroom (General Accountability Office, 2006).

A study commissioned by the U.S. Department of Education, *An Evaluation of Teachers Trained Through Different Routes to Certification*, found little difference in the outcomes achieved by students whose teachers were prepared through non-

Text Box 9.1 Attracting and retaining teachers.

The Proud School District

Among the fourteen school districts in this metropolitan area, there was one that prided itself on being innovative and cutting edge in all aspects of its operation. It was a growing school district with new schools being built every few years. The demographics of the district were represented by many families with professional backgrounds and economic affluence. Student achievement was among the highest in the state.

Historically, the school district did not have to aggressively recruit for new teachers. In many cases it was the first choice for teachers who wanted to work in the area. As the school board members, superintendent and personnel director often stated, "teachers want to work for the Proud School District." At the annual teacher recruitment fair at State U, the district would be flooded with applications. Meanwhile, the central city school districts would collect a fair share and the small rural districts would hope to get a few. Proud School District always had a ready pool of applicants to fill vacancies and to staff new schools.

The school board and superintendent were so confident of the attractiveness of their school district that they felt it was not necessary to pay attention to the salary schedules of their neighboring school districts. Several of these districts had been assertively moving to increase their salary and benefit packages in order to make their school districts more attractive to teachers looking for employment. Over time the pay differential between the Proud School District and many of its neighbors was substantial.

The problem hit home when the personnel director approached an outstanding student teacher who had been recommended by one of the district's principals for a position in her school. When offered a teaching contract in the district for the next year, the student teacher declined, indicating she had accepted a contract in a neighboring central city school district. The personnel director's response was surprise as he stated, "but everyone wants to work for the Proud School District."

The Proud School District eventually got its salary schedule back up to a competitive level, but this took many years to accomplish. By neglecting the teacher salary schedule, the school district created a problem that could not be fixed in a year or two. Millions of dollars had to be added to the schedule each year to catch up, and other district needs had to be put on hold until this was accomplished.

traditional means (Constantine, Player, Silva, Hallgren, Grider and Deke, 2009). The study compared elementary school student achievement in similar grades. In the evaluation, alternatively certified teachers were matched against traditionally certified teachers, and differences were mostly ascribed to normal variation. The report cited several other studies, which found similar results.

The Standards of a Profession

There are those who argue that teachers are more than well compensated and even go so far as to assert that teaching is not a profession like law or medicine, but rather a craft or trade. However, such assertions are ludicrous in the face of most definitions of a profession. The standards upon which a profession is judged clearly cover teaching. A profession involves many years of education or specialized training, and a bachelor's degree is the norm for minimum qualifications to teach.

Professionals have special knowledge, and it is common for those aspiring to teach to pass a state examination in which they must demonstrate special knowledge as a condition of licensure. Professionals are compensated with a salary, usually stated on an annual basis or as a fee, like doctors and lawyers in private practice. Most school districts use a "salary schedule" to determine how teachers will be compensated for the year.

Professionals are required to apply discretionary judgment and problem-solving skills in order to carry out their duties. Clearly, teachers apply such discretion on an ongoing basis as they plan and deliver instruction, and assess student work. Professionals are not supervised on a constant basis; teachers most often work independently and are supervised on a periodic basis. Professionals solve problems of practice and collaborate with other professionals to do so. Teachers work in a collegial environment focused on improving the school and student learning.

The U.S. Department of Labor publishes criteria for determining which employee groups are exempt from the overtime pay requirements under the Fair Labor Standards Act (FLSA). The act specifies that professional employees, as opposed to wage earners, are exempt from the provisions of the act. Here is how the U.S. Department of Labor defines such professionals:

> To qualify for the learned professional employee exemption, all of the following tests must be met:
>
> o The employee must be compensated on a salary or fee basis (as defined in the regulations) at a rate not less than $455 per week;
> o The employee's primary duty must be the performance of work requiring advanced knowledge, defined as work which is predominantly intellectual in character and which includes work requiring the consistent exercise of discretion and judgment;
> o The advanced knowledge must be in a field of science or learning;

o And, the advanced knowledge must be customarily acquired by a prolonged course of specialized intellectual instruction. (U.S. Department of Labor, 2012)

By modern standards, teaching ranks as a profession like any other. It should no longer be considered a part-time job or avocation for those waiting for a "real career" to develop. As professionals, educators should expect to be compensated like professionals.

Support Staff

School districts employ an array of workers beyond the classroom. Almost half of the people on a typical payroll are not in the classroom and many of these employees have no direct contact with students. These jobs can range from professional positions like accountants, architects and lawyers to craftspeople like carpenters, plumbers and electricians. Full- and part-time employees such as bus drivers, food service workers and aides of various types are also found among these non-certified or unlicensed staff. Wages for such employees are usually benchmarked to salaries for comparable positions in the private sector within the prevailing regional market.

Other licensed employees without classroom duties include therapists, psychologists, curriculum specialists and a host of others. These specially trained employees comprise the assortment of professional and other support personnel that make up a school district's workforce. In some states such employees are found in intermediate education entities like a Board of Cooperative Educational Services (BOCES) or intermediate school districts. These units can be funded through taxes based on a regional taxing structure, through cooperative arrangements in which area school districts join together to fund such entities or on a fee-for-service basis as a quasi-private enterprise. Various state laws and regulations allow for and control such arrangements.

Administrators

Compensation for management staff in education has historically been the focus of criticism and controversy. Perceived exorbitant pay packages are routinely reported by local news media in the form of exposés. School districts are also admonished as being bloated with redundant, overpaid administrators. But such hyperbole represents more myth and legend than fact. Compared to other government organizations and businesses, schools tend to be much more efficient and are run with fewer supervisory personnel. On average there is one school leader to supervise fifteen employees. This far exceeds other areas like banking, publishing, manufacturing and the military, which have a much lower ratio of employees to supervisors (McLane, 2008).

The Bureau of Labor Statistics (2009) reports that for the year 2007, the average school administrator salary was a bit over $82,000 per year. This figure masks a wide range of salary and benefit packages that exist among education leaders. Some assistant principals will earn salaries in the low $50,000 range, while some secondary school principals will earn $150,000 or more. Administrator salaries are highly correlated to the size or enrollment of the school or school district and regional cost-of-living indexes. Within a school district, building principals' salaries rise along with the grade level of the school, i.e., elementary, middle and high school, which in turn is accompanied by larger enrollments, larger staffs and more responsibilities. Suburban and urban school leaders tend to earn more than rural and small-town administrators. Variations among states are also broad: Connecticut, New Jersey and New York pay substantially more than Iowa and Montana. School leaders in private and charter schools earn about one-third less, on average, than their public school counterparts, even after factoring in school size.

Superintendent salaries have grown to significant levels in a small minority of school districts. About a decade and a half ago, top-end superintendent salaries broke the $200,000 barrier, and today we see a handful of cases where superintendents are making over $300,000. Most superintendents earn much less. Their salaries and other compensation are circumscribed by the same parameters that affect school principal salaries: size of the school district, size of the budget, number of employees and what other superintendents make in other like communities in the state. Their benefit packages most closely resemble what other employees in the school district get. Often, because superintendents do not earn tenure and have limited contract terms, they will be given a supplemental retirement annuity. Compared to chief executive officers (CEOs) in the private sector, school superintendents tend to earn much less on average. This difference tends to hold up even when comparing the number of employees in the company and school district or the size of the annual budgets. Salaries for city managers, as well as fire and police chiefs, are frequently the reference salaries school boards use for superintendents. However, these organizational leaders usually run smaller operations with fewer employees and smaller budgets and don't have the level of formal education expected of a school superintendent.

The Single Salary Schedule

One of the more criticized aspects of public education policy today is the ubiquitous single salary schedule. Sometimes called the uniform salary schedule, it is easily recognized by the steps and lanes that outline the annual salary for a teacher based on years of experience and levels of education; see Table 9.1 for an example. This method of determining teacher salaries is found in 95 percent of school districts today. Detractors complain that this method of compensation rewards mediocrity and reduces incentives for teachers to work hard, excel and innovate.

But as is often the case among education policy critics, they fail to understand, or choose to ignore, the reasons behind existing policy. They disregard the background behind the established practice. When first introduced in the early twentieth century, the uniform salary schedule was hailed as a major innovation and

Table 9.1 Teacher uniform salary schedule.

Education Level	BA	BA+16	BA+32	MA or BA+48	MA+16	MA+32	MA+48	MA+64	MA+80	MA+96 or Doctorate	Education Level
LANE	1	2	3	4	5	6	7	8	9	10	LANE
	Salary	Salary	Salary	Salary	Salary	Salary	Salary	Salary	Salary	Salary	
STEP											STEP
A				33581	34937	36293	37649	39005	40361	41717	A
B	31728	32225	33581	34937	36293	37649	39005	40361	41717	43073	B
C	32225	33581	34937	36293	37649	39005	40361	41717	43073	44429	C
D	33581	34937	36293	37649	39005	40361	41717	43073	44429	45785	D
E	34937	36293	37649	39005	40361	41717	43073	44429	45785	47141	E
F	36293	37649	39005	40361	41717	43073	44429	45785	47141	48497	F
G	37649	39005	40361	41717	43073	44429	45785	47141	48497	49853	G
H	39005	40361	41717	43073	44429	45785	47141	48497	49853	51209	H
I	40361	41717	43073	44429	45785	47141	48497	49853	51209	52565	I
J	41717	43073	44429	45785	47141	48497	49853	51209	52565	53921	J
K	43073	44429	45785	47141	48497	49853	51209	52565	53921	55277	K
L		45785	47141	48497	49853	51209	52565	53921	55277	56633	L
M			48497	49853	51209	52565	53921	55277	56633	57989	M
N				51209	52565	53921	55277	56633	57989	59345	N
O				52565	53921	55277	56633	57989	59345	60701	O
P				53921	55277	56633	57989	59345	60701	62057	P
Q					56633	57989	59345	60701	62057	63413	Q
R						59345	60701	62057	63413	64769	R
S							62057	64769	66125	67481	S
	Base Salary: $29016				Increment: $1356			General Hourly Rate: $23.79			
Note:	Placement on the Teacher Salary Schedule, for teachers new to Colorado Springs School District #11 shall be based upon										
	graduate-level semester credit hours only.										
Note:	Teachers new to the District may be granted up to 14 years of experience, provided the experience occurred within the last 15 years,										
	in accordance with the Master Agreement.										
Note:	Salaries listed above are for full-time(7 hours/day) employment and for a 185 day teaching contract. Teachers who work less than a										
	full-day (7 hours) and/or a full year (185 days) will have the amount listed above prorated.										

Source: Retrieved from http://www.d11.org/hr/Salaries/Teachers.pdf.

improvement of existing practice. The compensation method was established as part of the many major reforms that were taking place in government and education across the nation at that time. Furthermore, despite the ubiquitous use of the single salary schedule, half of the school districts that use it also use incentive programs to encourage faculty and staff to seek special training or to incentivize other activities deemed important to the school district.

Prior to the reforms of the Progressive Era, government, including many school systems, was characterized by political patronage, nepotism and cronyism. Political patronage is a system of rewards and sanctions doled out by elected officials. It is based on the idea that those who win an election should distribute government jobs and contracts to those people who helped get them elected and keep them in office. Common practice, prior to the introduction of a civil service system based on merit and qualifications, would see government workers, including teachers and administrators in some communities, thrown out of their government jobs upon the change of political officeholders like school boards.

Under that old system, every government employee from dogcatcher to chief of police was subject to losing his or her job when a new mayor took office in the town. Neither qualifications nor public service were as important as loyalty to a political party or ward boss. Corruption and government were synonymous terms in many parts of the country during that time. Millions of dollars of tax money were distributed through the patronage system. Nepotism was also a common feature of this system, where government largess was passed out to the family members of elected officials and political party bosses. Civil service, a system of government personnel management, was eventually adopted by most federal, state and municipal entities by the middle of the century.

Gender discrimination was yet another common practice in school districts at that time. Sometimes it came in the form of differential pay schemes for elementary and secondary teachers, with women dominating the elementary level and men the secondary. Men were routinely paid more than women even when they possessed the same level of education and experience. In some cases the discrepancy was as much as 60 percent less (Strachan, 2003). This type of discrimination also applied to building administrators, although women in administration were a small minority back then.

The uniform salary schedule was pioneered in urban school districts like Milwaukee and Denver (Seyfarth, 2008). It was seen as an innovative reform at the time. Rather than set salaries on an individual basis or based on gender, the new approach sought fairness and transparency. Experience and level of education were seen as valuable attributes for teaching. It was believed the organization gained from having educated and seasoned personnel. Remember, too, that many people entered teaching with limited education, so inducements to encourage faculty to continue their education were an important policy consideration. Today, many school districts offer financial incentives for teachers willing to achieve National Board Certification. Also, consider that in order to develop teaching as a profession, it was important to have people pursue teaching as a lifelong occupation and not temporary employment until marriage.

It should be no surprise that the uniform salary schedule has lasted almost a hundred years. It is viewed as fair, understandable and predictable. These qualities benefit many stakeholders. Teachers have a clear picture of what their compensation will look like over time, which can help them decide to apply at a particular school district or move on to somewhere else. Policy makers and administrators have a pay system to work with that allows them to forecast payroll needs well into the future. This is very important when considering that teacher salaries are such a big part of a school district's operating budget. Taxpayers can view an easy to understand compensation program and know what they are paying for. The criteria for advancement on the schedule are easy to understand, transparent and readily measurable, so teachers are clear about what is valued by the school district and can see how they are treated relative to their peers.

Alternative Compensation Systems

Over the past several decades, attempts have been made to pilot or substitute various compensation systems for the unified salary schedule. Many of these plans last a brief period of time, only to have school districts revert back to the former approach (Harty, Greiner and Ashford, 1994). Three common reasons for the failure of such systems are: 1) a lack of funding to provide meaningful financial incentives or to sustain the program; 2) an overly cumbersome measurement system used to determine salary increases or bonuses, which are also plagued by issues of validity and reliability; and 3) resistance from some teachers and administrators based on questions of fairness, transparency and economic incentives (Heneman, Milanowski and Kimball, 2007).

The alternative compensation programs, sometimes called strategic compensation, can be classified under several models. The typology can be broadly characterized as: career ladders; school-based incentive programs; individual incentive programs; and combination programs. With the exception of the career ladder, most of these programs are based on the merit pay systems found in business and industry. However, the term "merit pay" has not been well accepted among educators and has been replaced by the more benign sounding "performance pay." Table 9.2 provides a summary of these kinds of programs.

Despite the enthusiasm for performance pay programs among some school reformers, political leaders, policy makers and the public, only a relative handful of school districts have adopted or institutionalized this alternative to the uniform salary schedule. However, the popularity of performance pay among political leaders has gained much momentum since 2010, and a number of governors and state legislatures have mounted efforts to adopt such programs as statewide school policy.

Yet, among these school district policies and proposals, few have pure merit pay programs. Here are some more reasons why:

o It is easy for merit pay systems to become complex. Many important school objectives are hard to measure and this leads to selecting things that are easy to measure like test scores. The merit pay measurement system can be cumbersome to develop and maintain. Teachers and principals have more to do than keep score for the merit pay plan. In order to make the system fair, it is made simple and thus ineffective—but teaching is complicated.

o The old adage goes, "What gets measured gets done," but teachers do much more than prepare students for tests. Is the reverse then true, "what doesn't get measured doesn't get done?"

o People are motivated by money, but only to a point and only by threshold amounts. Most school districts cannot afford to pay big bonuses to substantial numbers of teachers. School districts cannot afford to fund such programs; thus, the program doesn't have the desired impact among a critical mass of teachers.

o There is a danger in setting numerical goals and quotas for employees without giving them a method for reaching those goals. Incentivizing outcomes without a means will often—very often—lead to distortions in the system (Deming, 1993). Think of all the cases of people who have been caught "cooking the books" in such systems, whether it be in business, government or education.

Table 9.2   Typology of alternative or strategic compensation systems.

Typology of Alternative Compensation Systems	
Career ladder	**Theory of action:** People are motivated by money. Teachers do not all possess the same skill and knowledge. Teachers need incentives to develop skill and knowledge. The experienced educator should contribute more than the novice.
	Method of implementation: Over time teachers move through a series of steps up the "ladder." They must demonstrate increased skill and knowledge to advance. Many career ladder programs also require that the teacher take on additional leadership responsibilities outside the classroom like mentoring new teachers or chairing a grade level, department or curriculum committee. The stages or steps on the career ladder follow a pattern, for example, probationary teacher, novice, professional and master teacher. Successful annual evaluations, additional formal education and professional development are usually components of the program. Teachers must progress to a specified level within a certain period of time and then have the option to continue to higher levels.
	Method of compensation: The school district salary schedule is divided into segments to represent each segment on the career ladder. Payment for longevity within a level is truncated after so many years, thus providing an incentive to move to the next level. Periodic cost of living raises are generally awarded across the board. Substantial increases in salary are achieved by moving up the ladder.
Individual performance	**Theory of action:** People are motivated by money. The single salary schedule rewards complacency.
	Method of implementation: Teachers earn bonuses or advancement on a salary schedule for demonstrating increases in student achievement. Student achievement targets are preset by policy or are negotiated. The crude form of such a system relies on standardized tests exclusively. This is a merit pay program.
	Method of compensation: One-time bonuses, salary increases or a combination are based on the academic achievement of students. Compensation is often awarded on a sliding scale linked to the amount of increase in student test scores.
School-based performance	**Theory of action:** People are motivated by money. Schools are collegial environments. Peer pressure is motivating.
	Method of implementation: Teachers within a school earn bonuses or advancement on a salary schedule for demonstrating increases in student achievement across grade levels, in the school as a whole or through other indicators like student attendance or reduced dropout rates. Student achievement targets are preset by policy or are negotiated. The crude form of such a system relies on standardized tests exclusively. This is a group form of a merit pay program.
	Method of compensation: One-time bonuses, salary increases or a combination are based on reaching established targets for the school. Compensation can be awarded on a sliding scale linked to the amount of the target reached, e.g., the proportion of increase in student test scores.
Mixed incentive plan	**Theory of action:** People are motivated by money. The single salary schedule rewards complacency. Teachers do important things that contribute to school improvement and student learning but are hard to quantify. Not all teachers, schools and students are the same and thus judgments should be made on a case by case basis.
	Method of implementation: Teachers earn bonuses or advancement on a salary schedule based on a menu of items from demonstrating increases in student achievement, to earning a master's degree, to leading a curriculum committee, to working in hard-to-staff schools, to having a license in a shortage area like mathematics, science or special education. Annual goals are set by policy or negotiated with the site administrator or a combination of both. The crude form of such a system relies on standardized tests exclusively to judge student achievement. Merit pay is a small part of this program.
	Method of compensation: Bonuses, salary increases or a combination are based on the negotiated annual plan. Compensation is often awarded on a sliding scale, within a range, and linked to the level and amount of goal achievement.

Few people argue about the incentive programs found in some other occupations. When a lawyer wins a big case with a big cash settlement, he or she gets a big bonus. When a professional athlete leads the team in scoring and to the playoffs, his or her incentive contract kicks in. If a salesperson exceeds the monthly quota, there is more money in the pay envelope.

Like so many policy ideas that are directed at public education by elected officials, policy makers, researchers, think tanks and reformers, merit pay schemes have attracted a lot of sound and fury, but little has been added to systemic improvement. There are many reasons why, and some of them are listed above. But perhaps the biggest reason this policy initiative does not work is that it fails to understand how people are motivated and why they earn what they earn (Kanter, 1987; Pink, 2009; Ramirez, 2001; Ramirez, 2011). A common example of teachers not being motivated by money is seen when they agree to coach a sports team or sponsor a club. The hourly rate for such undertakings is frequently less than minimum wage—much less!

Pay bonus targets that are hard to reach need to have substantial dollar amounts tied to them in order to motivate. As discussed in previous chapters, school districts commonly budget on an incremental basis—that is, they up their budgets a couple of percentage points here or there each year. They don't have large pools of discretionary money to draw on for bonuses. This puts the school district into a zero-sum game in which it must not fund something else in order to fund the bonuses. One option is to not give cost-of-living increases to those who do not earn merit pay, but this will lead to contradictory policy goals. Remember that the main objectives of a compensation system should be to attract and retain qualified personnel. An ambiguous system that promotes doubt about future earnings or is perceived as unfair will do neither.

Advocates of performance pay programs point to the Denver Public Schools ProComp program as an example of a successful pay for performance system. That program was developed over many years after having been piloted from 1999–2003 in more than a dozen schools. It is a mixed-incentive program, which offers several ways for teachers to earn incentive pay aside from raising test scores. The program is funded in part by an annual $25 million voter-approved mill levy increase, which is in addition to the amount the district's per pupil operating budget raises locally. ProComp was negotiated between the school district and the Denver Classroom Teachers Association. Participation in the program was voluntary for teachers who had been hired by the district prior to full adoption. ProComp has been modified several times since its formal adoption in 2004 (Denver Public Schools, 2009).

Another such program is the Teacher Advancement Program (TAP), which is also a mixed-incentive model. Like ProComp, TAP uses a combination of career ladder, professional growth, added responsibilities beyond teaching, increased student achievement and increased school achievement to calculate salary advancement and performance pay bonuses. Schools and districts adopting TAP tend to use supplemental funding, like grants from foundations or the federal government, to launch the program (Sawchuk, 2009).

Alternatives to the uniform salary schedule may have some value. But one has to ponder, if merit pay is such a wonderful idea, why isn't it routinely found in private schools? Private schools don't typically have labor unions and are not encumbered by civil service regulations. Private schools are free to make choices about how they compensate their teachers. Perhaps it is because the issues raised above are part of the reality of operating schools.

Benefits

The goal of attracting and retaining quality employees is at the heart of an organization's compensation system, and benefits have grown to become an important component. It is estimated that 20 to 35 percent of the total compensation earned by a school employee is awarded in the form of non-wage benefits. These benefits routinely include an array of insurances, such as health insurance, and a retirement plan. Over the years the benefit portion of the school employees' total compensation has become very complex. Different classifications of employees—i.e., teachers, administrators and support staff—may have a different combination of benefits. Finally, some districts may offer individuals within each employee group an array of benefit options from which to choose in order to allow for a more customized benefit plan for the employee (Rosenbloom, 2005).

An example of different benefit programs among employee groups within a school district is commonly found in the area of vacation time. Because conditions and terms of employment are unique for each group, vacation time is awarded in a distinctive way as well. Teachers typically do not earn vacation time, but are hired under contract to be in school during the academic year and for some extra preparation and service duty days. However, in order to accommodate the needs of these employees, some school districts also allow for a limited number of personal leave days so that the teacher can take care of things like closing on a house. School administrators have longer annual contracts, which in some school districts allow them to earn vacation time. This arrangement provides flexibility to the administrator and school district to better meet the needs of the employee and school district. Support staff, such as custodians and school secretaries, might also earn vacation time but might be restricted to using that time during certain weeks when school is not in session.

Sick leave is another benefit afforded some employees. Sick leave allows an employee to miss work for specified medical reasons, usually for a limited number of days each year, without loss of pay. Some school districts allow employees to accumulate their annual allotment of sick leave days up to a certain amount. There are examples of school districts that pay employees for unused sick leave days, sometimes on an annual basis or upon retirement. There are even cases where districts allow employees to contribute their individually allotted sick days to a sick leave bank from which other employees can draw. As with vacation days, using sick days as a benefit that can be accrued and traded for payment creates a financial liability or obligation on the books of the school district.

Health insurance is another benefit employees regularly earn. It is usually awarded to workers who are employed more than half time. Here again, various options apply. Such benefits may or may not include coverage for eye care such as glasses, dental care beyond routine exams and the coverage of family members. It is common today for the employee to contribute to the cost of this benefit, particularly that portion associated with family coverage. Health insurance has been a runaway expense for school districts, growing at a rate twice that of inflation over the past two decades. To put this figure in perspective, consider that health costs in 1960 accounted for only 5.1 percent of the gross domestic product (GDP). This percentage grew to 17 percent of GDP by 2006. Employer spending for health insurance increased an average of 13 percent per year over the past six decades (Employee Benefits Research Institute, 2012).

The array of benefits and their costs often escape the notice of the typical worker until such time as they are needed. Most of the benefits have developed over time, either as a program initiated by an employer or as a mandate from the government. Profit sharing and employee pensions are two examples of benefit programs voluntarily introduced by individual businesses. On the other hand, programs like worker compensation insurance, devised to cover the cost of a worker injured on the job, and unemployment insurance, for workers who are laid off, are government-mandated benefit programs. The NCES presents extensive data regarding the types of benefits offered by school districts across the nation.[1]

With the introduction of the social security system in 1935, the government took unprecedented action to start to build a social safety net for the elderly, sick and disabled in America. For the first time, millions of Americans were given access to a retirement program that was backed by the federal government. However, certain state and local government workers, like teachers and other school employees, are often exempt from contributing to social security because they are covered by similar systems that are funded and run locally or by the state. In some states, large school districts might run their own retirement fund for teachers and other school district employees. A common model is the state teacher retirement fund into which all school districts contribute. Another model has all public employees, state or local, covered under one program. There are examples where some employees, like substitute teachers and temporary workers, are covered under social security because they are ineligible for the state system.

A key distinction of state and local retirement programs is that they are overwhelmingly classified as "defined benefit" programs. Under such programs the employer, or the employer and employee, contribute a portion of salary to the retirement program each paycheck. When the employee reaches a certain age and after a predetermined number of years of service, the retired employee can receive payments at a predetermined amount. This is usually calculated on a scale that considers years of service and average earnings while employed.

By way of contrast, many private sector employees are most often covered by a "defined contribution" program, which does not guarantee a specified retirement payment. Under this arrangement the employer and employee typically contribute to the retirement fund, which is invested in the hopes of growing the investment over time. Unlike the defined benefit program, there is no guarantee of how much will be available for retirement. Because both types of programs rely on investing the retirement funds in order to have sufficient money to pay retirees, both programs are subject to economic conditions affected by the business cycle. However, the defined benefit program guarantees a retirement payment amount, which is backed by the state or local government, and ultimately underwritten by taxpayers if necessary.

Some big distinctions between defined benefit and defined contribution programs are the issues of transportability and inheritance. Government-run pension programs, which are typically defined benefit, are not very transportable for employees who leave the state (or the school district, in those cases where the retirement program is run at that level). If the employee leaves the participating employer, only limited options are available to him or her. Assuming the individual is vested in the retirement program, i.e., has contributed for a requisite number of years, he or she can leave his or her contribution in the program and start drawing a pension when he or she reaches retirement age; the individual can withdraw the portion of his or her contributions to the program, assuming that was how the plan was structured, and forgo any pension benefit that might have come to him or her; or the individual can move to an employer who is also participating in the same retirement system and continue adding years of service and contributions to the system. For those employees who are not yet vested in the system and leave their employment, all contributions made by the employer on behalf of the employee are lost.

State-run defined benefit programs routinely have a survivor benefit option, which passes the retirement payment to a surviving spouse. To gain this benefit, the retired employee must accept a reduced payment. When both the retiree and spouse are deceased the payments stop. By way of contrast, most defined contribution programs allow the employee to move their retirement fund to another employer regardless of the jurisdiction, so long as the new employer has such a benefit program. Contributions made by the employer and employee are typically retained by the worker. And the fund is part of the employee's estate and thus can be passed on to heirs.

Critics of public education and teachers generally point to what they term as "lavish" benefit offerings, like defined benefit retirement programs, as hidden compensation that is not reflected in the stated salary. They decry the lament about low teacher salaries as ingenuous carping by greedy educators. However, detailed analysis of the data shows that despite the benefit packages available to many teachers, their total compensation still lags behind professions requiring similar education and skills (Allegretto et al., 2008). Furthermore, the recession of recent years has precipitated rollbacks on many of these benefits or required employees to make greater contributions to maintain the same level.

The Role of Labor Unions

Teacher associations and employee bargaining units are also the subject of overstatement as either the bane of public education or the reason for the advancements of education in America. Reality is likely somewhere in between. The right to bargain collectively was granted to teachers in the 1960s. This right is governed by national and state labor laws. The legislation varies from state to state and ranges from prohibited to enabling. The scope of bargaining and the parameters of the

Picture 9.1   Teacher strikes were widespread in the 1970s.

negotiation process are also defined by statute in many places. Here again, as with benefits, efforts by some governors and state legislatures to diminish the bargaining power of teachers and other public employees have seen much more attention in recent years.

The effects of collective bargaining on salary, benefits and working conditions are also debated. Whether improvements to the teaching profession and for other school employees can be attributed to unions and collective bargaining is questioned (Moe, 2011). Data show that states with laws that allow teachers and other school employees to bargain often have higher salaries compared to those states that do not. However, as detailed in this chapter, all states and school districts have a vested interest in offering competitive salary and benefit packages that attract and retain quality employees. Negotiated teacher contracts tend to have an upward push on salaries within a region, even among school districts that do not have bargaining units (Loveless, 2000).

Summary

The teaching profession has developed over time along with the institution of public education. Today, teaching is accepted as a learned profession and entitled to the benefits and privileges afforded other professions. As school districts have grown in size and complexity, the number and types of employees found in these organizations have also grown.

State systems of education and school districts have an incentive to hire quality professionals and other employees and thus strive to develop compensation systems that attract and retain such individuals. The uniform salary schedule has been in existence for almost 100 years and retains its dominance among methods of compensating teachers and other professional employees. Benefit packages for school employees have developed into a complex and expensive part of compensation systems. This segment of the system has increased in cost at a runaway pace in recent decades.

School districts are experimenting with alternative systems of compensation in the hope of incentivizing employees in new ways. This area of innovation, sometimes called strategic compensation, needs much study. Advocacy and resistance to moving away from the uniform salary schedule is widespread. It remains to be seen whether changes can be institutionalized on a broad basis.

Note

1. http://nces.ed.gov/surveys/sass/tables/sass0708_2009320_d1s_04.asp.

References

Allegretto, S. A., Corcoran S. P., and Mishel, L. (2008). *The teaching penalty: Teacher pay losing ground.* Washington, DC: The Economic Policy Institute.

American Federation of Teachers (2008). *Survey and analysis of teacher salary trends, 2007.* Washington, DC: American Federation of Teachers.

Bureau of Labor Statistics (2009). *Occupational employment statistics.*Retrieved from http://www.bls.gov/oes/2007/may/oes119032.htm

Cable News Network (Producer). (2009, January 30). *Day of vindication for grandma as pay law signed* [Television broadcast]. Atlanta: Cable News Network.

Constantine, J., Player, D., Silva, T., Hallgren, K., Grider, M., and Deke, J. (2009). *An evaluation of teachers trained through different routes to certification, final report* (NCEE 2009-4043). Washington, DC: National Center for Education Evaluation and Regional Assistance, Institute of Education Sciences, U.S. Department of Education.

Deming, W. E. (1993). *The new economics for industry, government, education.* Cambridge, MA: MIT Center for Advanced Engineering Study.

Denver Public Schools (2009). *Welcome to teacher ProComp.* Retrieved from http://denverprocomp.dpsk12.org/about/overview

Employee Benefits Research Institute (2012). *EBRI databook on employee benefits.* Washington, DC: Employee Benefits Research Institute. Retrieved from http://www.ebri.org/publications/books/index.cfm?fa=databook

General Accountability Office (2006). *Troops to Teachers program brings more men and minorities to the teaching workforce, but education could improve management to enhance results.* Washington, DC: General Accountability Office.

Harty, H. P., Greiner, J. M., and Ashford, B. G. (1994). *Issue and case studies in teacher incentive plans* (2nd ed.). Washington, DC: The Urban Institute.

Heneman, H. G., III, Milanowski, A., and Kimball, S. (2007). *Teacher performance pay: Synthesis of plans, research, and guidelines for practice.* Policy Briefs, The Consortium for Policy Research in Education. Philadelphia, PA: The University of Pennsylvania.

Hoxby, C., and Leigh, A. (2005). Wage distortion: Why America's top female college graduates aren't teaching. *Education Next, 5*(2), 50–57.

Kaestle, C. F. (1983). *Pillars of the republic: Common schools and American society, 1780–1860.* New York: Hill and Wang.

Kanter, R. M. (1987). Attack on pay. *The Harvard Business Review, 65,* 60–67.

Loveless, T. (Ed.). (2000). *Conflicting mission?* Washington, DC: The Brookings Institution.

McLane, K. (2008). Countering three misconceptions about administrators. *The School Administrator, 65*(11).

Moe, T. (2011). *Special interest: Teachers unions and America's public schools.* Washington, DC: The Brookings Institution.

National Center for Education Statistics (2007). *Digest of education statistics.* Washington, DC: National Center for Education Statistics.

National Center for Education Statistics (2011). *Digest of education statistics.* Washington, DC: National Center for Education Statistics.

No Child Left Behind Act of 2001, Pub. L. No. 107-110.

Pink, D. (2009). *Drive: The surprising truth about what motivates us.* New York: Riverhead Books.

Ramirez, A. (2001). How merit pay undermines education. *Educational Leadership, 58*(5), 16–19.

Ramirez, A. (2011). Merit pay misfires. *Educational Leadership, 68*(4), 55–58.

Rosenbloom, J. S. (2005). *The handbook of employee benefits: Design, funding and administration* (6th ed.). New York, NY: McGraw-Hill.

Sawchuk, S. (2009). TAP: More than pay for performance. *Education Week, 29*(27), 25–27.

Seyfarth, J. T. (2008) *Human resources management for effective schools* (5th ed.). Boston: Allyn and Bacon.

Strachan, G. C. (2003). *Equal pay for equal work.* In Hoffman, N., *Women's "true" profession: Voices from the history of teaching* (p. 316–320). Cambridge, MA: Harvard Education Press.

Webb, L., Metha, A., and Jordan, K. (2007). *Foundations of American education.* Upper Saddle River, NJ: Merrill.

U.S. Department of Labor. (2012). The fair labor standards act. Retrieved from http://www.dol.gov/elaws/esa/flsa/overtime/p1.html.

Revenue for Schools   **10**

Aim of the Chapter

IN THIS CHAPTER AN OVERVIEW OF THE THEORIES, policies and methods of taxation is presented, along with an explanation of their purposes and functions. Special attention is paid to systems of taxation related to revenue streams for the support of the public schools. The context for taxation policy and systems of revenue collection for schools is presented. Policy issues of taxation associated with theories of equity and justice are examined. The interrelationship between economic theory and tax policy is also discussed.

Introduction

The combined revenue for school districts across the United States approaches $600 billion annually. This money is raised through a complex system of taxation that involves local government, state government and the federal government. Large sums of money are collected and transferred among the different governmental entities and ultimately end up paying for everything from pencils to teacher salaries.

Taxes in various forms have been a part of human existence since the beginning of civilization. They are, as Benjamin Franklin reminded us, one of the two inevitable and unavoidable things in life, "death and taxes" (Notable Quotes, 2012). Even the Christmas story has a tax connection; Joseph and Mary trekked to Bethlehem in response to an edict from the Roman authorities in order to be counted in the census in preparation of the tax rolls. The founding of the United States sprang from a tax revolt against the English. The Boston Tea Party, a protest against unfair taxes, is a part of American history and folklore taught to all school children.

Taxes and tax protests did not vanish with the end of the American Revolution. No sooner had the nation been founded than President George Washington had to take to the field, again, with the Army to put down the "Whiskey Rebellion," an armed tax revolt. Even in modern times tax protests take place as one group or another perceives itself to bear an unfair burden. California's Proposition 13, approved by voters in 1978, was the prototype for similar residential property owner uprisings in a number of other states in the 1980s and 1990s. Today we see the Tea Party as the latest manifestation of an anti-tax movement. In all of these cases, extant economic conditions combined with the government's need to raise revenue precipitated such taxpayer unrest.

Within each state the schools are typically the biggest consumers of tax revenue from their respective state budgets. On average, one-third of state and local revenues go toward the support of schools. The PK–12 system of education commonly takes the largest share of state and local revenue, more than higher education, police, prisons and welfare. Multiple forms of taxation from different levels of government are applied in order to raise the revenue needed to fund the schools.

Funding for schools in many of the states has doubled over the past several decades, which has put further pressure on states to find needed money. Even in states with little enrollment growth per pupil, spending has increased substantially (Hoo, Murray, and Ruben, 2006). While there are several customary methods of raising revenue for schools, each state has its own system of taxation with particular nuances and uniqueness.

Taxation Policy and Theories

A tax is a resource, usually money, collected by government for its support. Taxes are compulsory and collected from people and businesses in many direct and indirect forms. Today, taxes are collected to accomplish many and varied government functions. Yet some of these government purposes are as old as civilization itself and are rooted in the very idea of government. Consider that the main and most important role of a national government is to protect its citizens from harm by foreign powers. The need to raise revenue to provide for the defense of a nation has been a challenge to sovereigns and governments throughout history. Levying a tax upon one's countrymen to equip and support an army and navy is a familiar theme in history. This kind of financial need by the government is generally recognized by a nation's inhabitants as necessary, particularly when an imminent threat exists.

As communities organize at the local level, they quickly recognize the need for a government role to maintain the peace within the population. Thus, police, court systems and jails are created, which require funding to be sustained. Once basic safety is provided, broader government purposes are attended to as a nation and its communities progress. The modern developed nation state of today is characterized by large, complex government systems whose roles cover everything from promoting commerce to ensuring the welfare of the old. Within this mix of government functions and taxation, schools have come to take on an important place.

Sovereigns and governments throughout time have struggled to craft taxing systems that produce needed revenue while avoiding the kinds of backlash from those being taxed that create even bigger problems for the taxing authority, e.g., a revolt. The legend of Lady Godiva of the 10th century, who implored her husband to remove an onerous tax from their subjects, comes to mind. From such struggles, certain policy theories of taxation have come to light. Within this theoretical realm are well-established principles of taxation that have come to be generally accepted as expected and workable policies. Adam Smith, the 18th-century economist who is famous as the father of free market economic theory, propounded a series of maxims regarding taxation. His seminal work, *An Inquiry into the Nature and Causes of the Wealth of Nations*, was first published in 1776, and he devotes a substantial portion of the book to the question of taxation policy and government revenue. Below is a paraphrasing of Smith's four maxims:

1. Everyone should contribute to support of the government in proportion to their ability to pay. It is important that a taxing system be fair, equitable or just.
2. A tax should not be arbitrary or unpredictable. It should be clear to all how much is due, when it is due and in what manner it is to be collected.
3. Taxes should be levied in a manner that makes their collection convenient for the taxpayer.
4. The tax itself should be efficient in that it takes only what is necessary. It should not be costly to collect and should not cost the taxpayer to pay. A tax should not punish industriousness on behalf of the people or encourage circumvention of the law, e.g., smuggling. And tax collection should not turn the government into an oppressor of the people.

Smith's maxims made good sense in eighteenth-century England and have much applicability today. They revolve around a core of fairness, practicality, efficiency and common sense. Most people recognize the need for a taxing system and pay their share with little complaint, but when the balance is lost within the system, problems surface.

Within the overarching concept of fairness in taxation are several underpinning principles, which devolve from these theories. Fairness, or equity, in taxation is commonly viewed along two perspectives—horizontal and vertical. Horizontal equity maintains that fairness is achieved when the tax system treats people in similar circumstances in an equal manner. By contrast, then, vertical equity asserts that equity is retained when folks in different economic situations are treated differently.

Modern tax system policy often strives to preserve equity through structuring its objectives to achieve both horizontal and vertical equity. This equity goal is most readily attained through a progressive tax configuration designed to raise the tax burden as the object or amount being taxed—for example, income—increases. A regressive taxing scheme would accomplish the opposite. With a regressive taxing method, the burden of taxes increases for the less wealthy and diminishes as wealth increases. In some situations property taxes can be considered regressive, for example, when two homes are assessed at the same rate for tax purposes and one is owned by a retired couple on a modest fixed income while the other is owned by a high-income family.

A third approach to taxing is the proportional tax, which sets a uniform rate of taxation without regard to the value of the thing being taxed or the wealth of the person paying the tax. Consumption and sales taxes are usually crafted as a proportional tax, where each consumer pays the same rate of taxation in proportion to how much the individual uses or purchases, regardless of their personal wealth. In consideration of the regressive potential of these taxes, some states exempt food and medicine from the levy.

There is an old adage about taxes by the late Senator Russell B. Long, which at once recognizes the need for taxes while also expressing a common sentiment: it goes, "Don't tax me. Don't tax thee. Tax the one behind the tree" (Brainy Quote, 2012). Policy makers who struggle to secure revenue for government purposes are constantly faced with the question of who should be taxed. Even when much thought and effort are put into designing a tax policy, the issue of who ultimately bears the burden is often elusive.

There is a phenomenon in taxing called "tax shifting," which compels the individual or business to get out from under a tax. For example, if rental property is taxed, then landlords are targeted to pay the tax. But in most cases the landlord shifts the burden to the renter in the form of higher rents. When there is an option to pass a tax along to someone else, the taxpayer will shift the burden.

But tax shifting is not unlimited. Economic concerns with supply and demand also come into play. So long as the thing being taxed retains a high demand, as in the example of rental property, then the tax shifting will hold up. But at some point the tax burden can become so high that the consumer, i.e., renter, will change

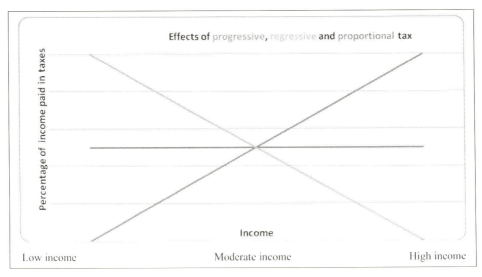

Figure 10.1   Results of various taxing systems.

behavior to avoid the tax. Some options in this case might be to buy a home instead of rent or get a roommate to share expenses. When a commodity or service is taxed too high, so that the overall price affects demand, consumers will look for substitutes if they are available. In some instances tax shifting can lead to a net loss of revenue. Tax policy that adversely affects the supply/demand equation will affect net tax revenue as well.

Governments will use tax policy to stimulate economic growth with the objective in mind of increasing net tax revenue. Examples of such policies are seen when state or local governments give tax breaks to businesses, for example, a municipality exempting a company from property tax if it locates a new industrial plant in the city. The strategy behind such arrangements is that other economic activity—such as increased income, new residential construction for those employed in the plant or increased retail activity in the community—will add to the net revenue of the municipal government.

Governments frequently use tax policy to shape the public's behavior; taxes on cigarettes and other tobacco products are examples. In this case the government intentionally puts a high tax burden on the commodity in order to stifle demand. By way of contrast, governments will use tax policy to increase demand for something, for example, when it gives a tax rebate or credit for consumers who buy fuel-efficient cars. Consequently, government tax policy can have aims beyond solely collecting revenue. This often makes tax policy complex, and because of things like tax shifting, it is frequently difficult for government to predict the impact of a tax.

Typology of Taxation

Governments can generate revenue from many sources other than taxes: land sales, leases for the exploitation of natural resources, fees for service and licenses are examples. However, taxes are by far what modern governments rely on for revenue to fund their operations. In the early years of the republic, the national government got more than 50 percent of its revenue from import taxes, but today that represents less than 1 percent. Income taxes first appeared on the scene during the American Civil War with rates that ranged from 5 to 10 percent. This raised $200 million for the Union in 1865. A national sales tax, actually a commodities tax, was placed on coffee, tea and sugar as well. War bonds were also sold and raised $2.5 billion for a federal government that was spending a million dollars a day on the war (Cox-Richardson, 2007).

In the United States the federal government gets 90 percent of its revenue from income taxes: specifically, personal income tax, corporate income tax and payroll taxes (U.S. Department of the Treasury, 2008). The states rely more on sales tax, and depending on the state, income, and property taxes (National Association of State Budget Officials, 2012). Local governments tend to depend on property tax, and to a lesser extent, fees for services, sales tax and payroll taxes.

Table 10.1   Federal income tax rates then and now.

Civil War	$600–5,000 5%	$5,000–10,000 7.5%	Over $10,000 10%	
Year 2011	Up to $17,500 10%	$70,700–142,700 25%	$217,450–388,350 33%	Over $379,150 35%

Each state designs its own taxing system, and as a result, some states—Nevada, Texas, Washington and Wyoming, for example—do not have a state income tax. Sales taxes will be of varying amounts among the states and even differ among governmental jurisdictions within a state. The following is a list of the more routinely found significant taxing methods used by governmental entities in the United States:

Personal Income Tax

This is a tax on the money individuals earn from employment salaries, rent on property they might own, dividends from investments or the "capital gains" from the profits from the sale of an investment, for example from a stock or property. Personal income tax is generally recognized as a progressive tax because it reflects a person's ability to pay. The payroll tax is an example of an income tax applied to salary and wages.

Corporate Income Tax

This is a tax on the profit made by a corporation. Economists often criticize this tax for several reasons: complicated tax codes frequently obscure the differences between profit and allowable expenses that can be shielded from being taxed; individual shareholders are taxed on dividends (based on company profits) they receive from the corporation, causing economists to point to this as a double tax; corporations use tax shifting when available; and corporate income taxes burden businesses and are said to diminish the economic viability of the nation.

Consumption Taxes

These taxes come in many forms. A general sales tax is applied at the point of sale for things bought and sold. In some states this also includes many retail services that are purchased directly by consumers, like haircuts and massages. Because of the tendency for it to be a regressive tax, it is common to see items like food, medicine and clothing exempt from a general sales tax. An excise tax is a sales tax on a specific commodity, for example, rubber, or gasoline. The ad valorem, or value added, tax is a tax on the additional value captured at the point of transaction, for example, a wine grower sells grapes to a vintner, the winemaker sells wine to a wine wholesaler, who then sells to wine retailers. State and local governments rely on consumption tax revenue sources extensively.

Severance Tax

This type of tax is applied to natural resources extracted from the earth like oil, gas, mineral, timber, fish, etc. The tax is levied against the value of the material. Some states, such as Alaska, have an abundance of such natural resources, while other states have very few of these resources to tax.

Motor Fuels Tax

This is an excise tax and deserves special mention because it is such a big revenue source for the federal government and states. Much of this money is earmarked for the nation's transportation infrastructure.

Cigarette, Tobacco and Alcohol Taxes

These are examples of an excise tax and are also classified as a sumptuary tax, otherwise known as a "sin" tax. Such taxes are designed to shape consumer behavior and stem from a tax policy philosophy that is broader than mere revenue collection. Policy makers have historically found it easier to establish sin taxes. However, as explained above, supply and demand dynamics along with tax shifting tend to make sumptuary taxes a less valuable revenue resource, because as the heavier tax raises the price of the commodity, demand is lowered.

Gambling

Some question whether state-sponsored gambling is a tax, since it is a voluntary activity. Yet gambling has been a source of government revenue since ancient times. Archaeologists regularly find gaming artifacts in their excavations. The United States is no exception, and its history is full of lore and legend about gambling and gamblers from riverboats to old west saloons (Dunstan, 1997). Today, government in the United States is the biggest gambling promoter in the world. Proceeds from gambling are an important source of state funding; for example, 37 states and the District of Columbia collect about $50 billion annually in proceeds from state-sponsored lotteries (Hansen, 2004). The many forms of gambling sponsored, taxed and licensed by states is quite remarkable. Even parochial Iowa offers more forms of gambling than Las Vegas, Nevada. While a significant contributor to state revenues, total proceeds from gambling are only a small portion of the $1.3 trillion collected annually by state and local governments.

Fees

Fees for services are routinely charged by governments as a revenue source. In some cases such fees are established as revenue neutral and are put in place merely to offset the governments' cost for providing the service, for example, building inspection for new construction or fishing license fees used to protect wildlife habitat.

Fees are often criticized when the amount of the fee exceeds the cost of providing the service. In such cases, fees are often called a hidden tax.

Property tax

This tax overwhelmingly represents the principal tax source for local governments. The federal government does not collect property tax. A property tax is most commonly made (levied) against land, buildings and larger personal property like cars, boats, farm equipment and industrial machinery. It is also called an ad valorem tax, which refers to the value added to raw land. Property taxes are regularly criticized as being regressive because they are regularly assessed without regard to the property owner's personal wealth, or in the case of agricultural, commercial and industrial property, the profitability associated with the property. It is common for different types of property to be taxed at different rates even when they have the same assessed value or market value, for example, a power plant, and a gold mine. Arcane taxing policies result as governments try to mitigate the regressive aspects of property taxes with exceptions, exemptions and rebates. Government methods for assessing, i.e., establishing the value of property for tax purposes, are also a complicated and controversial part of the process. Property tax systems vary greatly among the states.

Tables 10.2 and 10.3 depict estimates of federal and state revenue by principal sources. Note the significant proportion income taxes contribute to the federal budget. By way of contrast, see how import sales and property taxes contribute to state and local government resources.

The assorted revenue streams for government are rooted in tax policies that are based on an understanding of economic theories, which tend to hold up over time. It is acknowledged that the flow of wealth within a modern nation is found in the income of the populous, whereas the stock of wealth is contained within the property held by its citizens. Government tax policies reflect this economic reality in the form of property, consumption and income taxes.

Table 10.2   Estimated federal government revenue FY 11.

United States Federal Government Revenue Fiscal Year 2011 (data modified from original sources; amounts in billions of dollars)	
Income Taxes	1,598
Social Insurance Taxes	807
Excise and Sales Taxes	74
Fees and Charges	0.1
Business and Other Revenue	0.1
Balance Forward	85
Total Revenue	2,174

(Gross domestic product is estimated to be $14.7 trillion for this fiscal year, 2011).
Source: U.S. Office of Management and Budget.

Table 10.3   Total state and local tax revenue by source 2008.

Total	Personal income	Corporate income	Individual property	General sales and gross receipts	Motor fuel sales	Tobacco product sales	Alcoholic beverage sales	Motor vehicle and operator's licenses	Other
$1,285,566	300,740	56,317	404,466	304,772	38,495	16,551	5,633	23,171	135,421

Twelve months ending June 2008 (figures in billions of dollars)

Source: U.S. Census Bureau, 2008.

Picture 10.1   The valleys and peaks of the business cycle affect school funding.

However, numerous economic conditions also affect the tax policies designed by governments. Market conditions in the form of the business cycle and inflation and deflation in the monetary system are examples. As has been displayed in this chapter, government reliance on sales and income taxes produces vast amounts of revenue. But these revenue streams are increased or diminished as the income of the tax-payers increases or declines. This movement is based on the business cycle, which reflects the expansion and contraction of the economy. Business cycles exist at all economic levels: global, national, regional, state and local. They are affected by innumerable economic factors, from international monetary policy to a local employer going out of business. The recession of 2008 and its slow recovery greatly diminished tax revenue for all levels of government. As a result, there were major government cutbacks and layoffs of many government employees such as teachers and other school personnel.

Market economists contend that government tax policy can affect the business cycle. They argue that government taxation during an economic downturn can retard recovery, thus prolonging a recession. Even though tax revenues are in decline, market economists advocate for tax cuts as a means of stimulating the economy and economic growth during periods of economic contraction. The theory asserts that overall tax revenue will increase as personal and corporate incomes rise. The net result, according to the theory, is an increase in government tax revenue. Income and consumption taxes trend in a corresponding fashion; when incomes are rising, consumption goes up, and the reverse is true when incomes are in decline. More or less taxes are collected from these sources in relation to the business cycle.

Historically property taxes, on the other hand, tend to be more stable from year to year, while subject to bigger cyclical movements in the economy. For example, during the 1970s inflation in the United States spurred a rapid increase in the value of real property. Individual homeowners saw substantive increases in the value of their homes, as did businesses for the commercial property they held. As a result, local governments saw dramatic increases in their revenue even as tax rates remained constant.

Unfortunately, this inflationary movement correlated with a rapid increase in the cost of living, which was not matched by income growth. This resulted in property owners being squeezed by the government to pay what became exorbitant property taxes, particularly on residential property, which does not generate income. Many state and local tax policies were revised to reflect these economic conditions.

More recently, the value of property nosedived across the country after a period of heated and rapid price growth during the preceding decades. The so-called "housing bubble" burst and resulted in a steep decline in property values not seen since the Great Depression of the 1930s. This combined with an overall decline in the economy, or a recession, which saw considerable diminution of personal and corporate income. Record foreclosures on personal and commercial property along with large-scale unemployment resulted in an economic malaise. Individual and government revenue resources declined in a downward spiral as one condition affected the other.

Thus, the business cycle, the peaks and troughs of the economy over time, correlates to the revenue streams for governments as well as schools. Furthermore, government can affect the economy in big and small ways through its taxing policy. Careful planning and thoughtful policies are essential to craft a tax system that is fair and renders sufficient resources for needed government services. This is a very difficult task for policy makers, who must operate in a high-pressure and highly political environment.

Revenue for Schools

On average across the country, the state general fund is the largest single revenue source for schools. Table 10.4 depicts a breakdown of the origin of sources of

Table 10.4   Percentage distribution of revenues for elementary and secondary schools by source and state FY 2009.

State or jurisdiction	Total	Local[1]	State	Federal
		Revenues (in thousands of dollars)		
United States[2]	$593,061,181	$259,250,999	$277,079,518	$56,730,664
Alabama	7,239,083	2,295,475	4,166,018	777,591
Alaska	2,262,964	488,356	1,459,658	314,949
Arizona	9,771,972	4,040,008	4,594,648	1,137,316
Arkansas	4,823,956	1,583,147	2,684,309	556,500
California	70,687,012	20,895,829	40,605,913	9,185,270
Colorado	8,353,849	4,105,376	3,670,240	578,233
Connecticut	9,871,755	5,588,751	3,842,177	440,826
Delaware	1,755,133	517,796	1,094,909	142,428
District of Columbia[3]	1,651,014	1,475,283	†	175,732
Florida	26,322,090	14,579,923	9,047,588	2,694,579
Georgia	18,017,477	8,548,478	7,780,725	1,688,274
Hawaii[3]	2,689,757	91,889	2,205,032	392,837
Idaho	2,243,784	504,812	1,509,815	229,156
Illinois	26,512,711	16,041,221	7,324,750	3,146,741
Indiana	12,569,782	6,172,042	4,964,928	1,432,813
Iowa	5,519,854	2,530,666	2,545,360	443,827
Kansas	5,757,927	1,980,973	3,323,346	453,608
Kentucky	6,641,128	2,107,627	3,802,150	731,351
Louisiana	8,099,981	3,095,662	3,740,262	1,264,057
Maine	2,575,516	1,202,765	1,127,032	245,719
Maryland	13,097,508	6,703,926	5,698,735	694,847
Massachusetts	15,102,480	7,790,028	6,036,202	1,276,250
Michigan	19,585,635	6,427,004	10,904,987	2,253,644
Minnesota	10,542,303	2,995,407	6,914,839	632,057
Mississippi	4,360,702	1,350,375	2,334,355	675,972
Missouri	10,042,753	5,783,128	3,425,716	833,909
Montana	1,595,197	622,089	774,091	199,017
Nebraska	3,455,794	1,961,810	1,213,317	280,666
Nevada	4,450,741	2,654,134	1,362,123	434,484
New Hampshire	2,717,115	1,566,547	1,003,249	147,318
New Jersey	25,283,290	13,717,006	10,525,550	1,040,733
New Mexico	3,820,116	575,152	2,675,916	569,047
New York	55,558,190	26,991,217	25,346,556	3,220,417
North Carolina	13,322,946	3,515,648	8,401,249	1,406,049
North Dakota	1,102,479	532,990	408,004	161,484

State or jurisdiction	Revenues (in thousands of dollars)			
	Total	Local[1]	State	Federal
Ohio	22,956,215	10,352,625	10,917,974	1,685,617
Oklahoma	5,729,610	1,916,378	3,042,487	770,745
Oregon	6,145,206	2,357,357	3,117,303	670,547
Pennsylvania	25,632,072	13,843,699	9,920,340	1,868,034
Rhode Island	2,232,149	1,199,044	817,590	215,514
South Carolina	7,702,962	3,260,758	3,679,907	762,297
South Dakota	1,241,892	628,359	410,179	203,354
Tennessee	8,283,928	3,539,325	3,809,467	935,135
Texas	46,962,119	21,974,171	19,973,129	5,014,820
Utah	4,542,690	1,589,970	2,387,698	565,022
Vermont	1,571,006	121,922	1,346,300	102,785
Virginia	14,964,444	7,746,272	6,303,648	914,524
Washington	11,903,510	3,371,667	7,146,394	1,385,449
West Virginia	3,281,385	976,347	1,938,999	366,038
Wisconsin	10,832,105	4,720,471	4,809,185	1,302,449
Wyoming	1,675,896	620,095	945,167	110,634
American Samoa	79,922	209	11,282	68,432
Guam	262,823	212,652	†	50,170
Commonwealth of the Northern Mariana Islands	65,538	225	34,602	30,711
Puerto Rico	3,542,658	3,787	2,462,725	1,076,147
U.S. Virgin Islands	243,079	203,042	†	40,037

† Not applicable.

[1] Local revenues include intermediate revenues from education agencies with fundraising capabilities that operate between the state and local government levels.

[2] U.S. totals include the 50 states and the District of Columbia.

[3] Both the District of Columbia and Hawaii have only one school district each; therefore, neither is comparable to other states. Local revenues in Hawaii consist almost entirely of student fees and charges for services, such as food services, summer school and student activities.

[4] Reported state revenue data are revenues received from the central government of the jurisdiction.

Note: Detail may not sum to totals because of rounding.
Source: National Center for Education Statistics, Common Core of Data (CCD), "National Public Education Financial Survey (NPEFS)," fiscal year 2009, Version 1a.

funding for school districts. These major flows of revenue come from local, state and federal governments. In all cases the federal government is by far the least significant funding source. However, according to these data, only seventeen states and the District of Columbia get most of their funding from local sources. Note that Hawaii and the District of Columbia are one school district systems, in which Hawaii gets 89.9 percent of its money from the state and the District of Columbia gets 87.8 percent of its money locally. These two distinctions about state and local funding are arguable and should be discounted as outliers when computing national averages.

Chapters 4, 5 and 6 examined reasons for the trend toward increased state-level funding for the public schools, most of which revolved around issues of equity for students or taxpayers. Over the past century policy leaders have made efforts to design a more equitable distribution of resources for school districts across their respective state. The result of litigation or the threat of litigation has led to more state-level funding in order to mitigate illegal disparities among school districts. And in some states, a shifting local tax base has required more resources from the state level.

It is important to remember that school districts are a creation of their state, shaped by the state legislature and overseen by the state's executive branch, most often the state board of education and state education agency. This is why Florida has 67 school districts and New Hampshire has 177. The logic and reasons for such diversity around the nation are rooted in the historical development and political vagaries of the individual state. Each state has determined its system for organizing and delivering public education. As part of the system fashioned within each state, a means of funding the schools has also been put in place. As a result one can see an array of funding methods and approaches nuanced by esoteric taxing policies and practices.

One historical artifact that remains in use today is the application of the terms "independent" and "dependent" school districts. Fiscally independent school districts are entities that have the capacity to levy taxes, whereas dependent school districts must rely on another governmental entity—for example, a city or county government—to approve their budget and levy taxes on their behalf. According to a publication from the Education Commission of the States (2004), 34 states have no fiscally dependent school districts; nine states have no fiscally independent school districts. Six states have both fiscally dependent and fiscally independent school districts.

The notion that a school district is truly independent is misleading. With the exception of a few cases that have state constitutional authority, taxing power for school districts comes from the state legislature. Even so-called independent school districts levy taxes within parameters set by the state legislature or state constitution, or are subject to local voter approval of their budget or for any tax increases. The Education Commission of the States report also indicates that thirty-five states have limitations on the amount of taxation a school district can levy, known as "tax caps."

Additionally, twelve states restrict revenue collection by constraining how much a school district budget can grow from year to year. This is called a "spending cap." An assortment of tax caps are used, for example, setting maximum allowable mill levies on assessed property, requiring voter approval for the district to move to another tier of taxation and tying property tax rates to income. Examples of spending caps are seen in the form of percentage increase limits linked to the rate of inflation and growth in enrollment, budget growth rates set by the legislature and voter approval to increase the school district budget.

Twenty-six states use other sources of revenue in addition to property tax; for instance, sales tax, motor vehicle tax, excise taxes on hotel rooms, mineral taxes or a tax on timber harvesting. However, overwhelmingly, property tax is the main source of local revenue for school districts in most states.

Text Box 10.1 Mill levy explanation, definitions and calculations.

Property Tax Calculation
Governments often use the term "mill" in regards to taxation, particularly with respect to property tax. The American Heritage Dictionary (1996) defines a mill as, "a monetary unit equal to 1/1,000 of a dollar." It is derived from the Latin word *millésimus* meaning thousandth. In English we commonly see similar Latin words like millimeter and millennium that refer to a unit of one thousand.

The word "millage" is used to convey a tax rate or levy stated in mills. Numerically a mill can be expressed in several ways: 1 mill = 1/1,000; .001; $1 per $1,000.

Millages are usually applied to the "assessed value" of a property. Assessed value is the figure used by government for tax purposes. Assessed value is usually, but not always, tied to "market value." Market value is determined based on what someone will pay for something, in this case a piece of property. Below are two examples of a millage applied to residential property.

Example A	Example B
Market Value = $250, 000	Market Value = $250, 000
Assessed Value set by the state at 10% of market value = $25,000 assessed value	Assessed Value set by the state at 30% of market value = $75,000 assessed value
Established Millage for support of local school district = 45 mills	Established Millage for support of local school district = 15 mills
Property tax to support local school district is $25,000 x .045 = $1,125	Property tax to support local school district is $75,000 x .015 = $1,125

Property tax is used to support many local government entities from city government to fire protection districts, to libraries. Therefore the total millage and tax bill for an individual property owner will be much higher than in the examples above. How the taxing system is configured in the state will determine how much the property owner will pay to each entity. Some states use the term tax per thousand dollars to express millage.

Summary

Taxes are an unavoidable fact of life that serves as a foundational element of society. As nations have developed into the contemporary modern state, taxing systems and policies have become ever more sophisticated and complex. A key aspect of any tax policy is the consideration of equity for the taxpayer. Taxing systems are affected by and affect economic conditions from the community to the nation level.

The public schools are the largest "tax eaters." On average they consume one-third of state and locally generated revenue. Funding for schools has risen by all measures since the establishment of public education in the nineteenth century. Property taxes have historically been and remain a significant source of local school funding. Today, the majority of school money, on average across the nation, comes from the state level, which tends to rely more on sales and income taxes.

Education policy leaders and school administrators understand the burden placed on the community for the support of schools. A key responsibility for these leaders is the stewardship of the resources they are provided. The wise use of tax dollars and the delivery of an outstanding education for their students is a smart strategy for maintaining community support.

References

American heritage dictionary of the English language, (3rd ed.). (1996) New York: Houghton Mifflin.

Brainy Quote. (2012). Retrieved from www.brainyquote.com/quotes/r/russellb1101810.html.

Cox-Richardson, H. (2007). *West from Appomattox*. New Haven, CT: Yale University Press.

Dunstan, R. (1997). II. History of gambling in the United States. In *Gambling in California* (section II). California Research Bureau. Retrieved from http://www.library.ca.gov/crb/97/03/Chapt2.html.

Education Commission of the States (2004). *Taxation and spending policies.* Denver, CO: Education Commission of the States.

Hansen, A. (October 2004). Lotteries and state fiscal policy. *Background paper, 46.* Washington, DC: Tax Foundation. Retrieved from http://www.taxfoundation.org/publications/show/65.html.

Hoo, S., Murray, S., and Ruben, K. (2006). *Education spending and changing revenue sources.* The Urban Institute. Retrieved from www.urban.org.

National Association of State Budget Officials, 2012. Retrieved from www.nasbo.orglsites/default/files/Spring%202011%20fiscal%20survey_1.pdf.

National Center for Education Statistics (2008). *Revenues and expenditures for public elementary and secondary education: School year 2005–06 (fiscal year 2006).* (NCES 2008-328). Washington, DC: National Center for Education Statistics.

Notable Quotes. (2012). Retrieved from www.notablequotes.com/f/franklin_benjamin.html.

Smith, A.(1776). The wealth of nations. In Cannan, E. (Ed.), *An inquiry into the nature and causes of the wealth of nations* (5th ed.) (1904). London: Methuen & Co. Retrieved from www.econlib.org/library/smith/SMWN.html.

U.S. Department of the Treasury (2008). *Internal Revenue Service data book 2006.* Washington, DC: U.S. Department of the Treasury.

Understanding Budgets   11

Aim of the Chapter

THE PURPOSE OF THIS CHAPTER IS TO DEMYSTIFY the budget process. Strategic elements of budgets and the budgeting process are covered. Key terms and concepts are explained, and recommendations for good budget practices are offered. The aim of the chapter is not to qualify you for your own set of green eyeshades and sleeve garters, but to help you gain insight into this important area of education finance, along with some basic knowledge and key terminology.

Introduction

For many individuals, the idea of a discussion about budgets or the budgeting process invokes images of nineteenth-century accounting clerks in their green eyeshades and sleeve garters hunched over tables in a counting house—think Mr. Marley and Scrooge from Charles Dickens' *A Christmas Carol* (1843). Unfortunately, there isn't much that can be done to spice up the topic, despite the fact that most education leaders and policy makers understand its central importance.

Among the top reasons superintendents lose their job is a loss of confidence by the school board in the superintendent's ability to handle the school district budget. School board members understand that as trustees of the community's resources, they must ensure that tax dollars are spent wisely. Most board members are not budget experts, but they learn to deal with the fundamentals of the school district's budget. Similarly, legislators covet seats on the budget committees, not because they love to study accounts, but because they understand that this is where real power resides in the legislature.

Despite understanding the importance of budgets, many educational leaders and policy makers shy away from studying the topic. Part of the reason for this is that they find the matter boring. And for many, the subject is intimidating because they fear it will involve numbers and complicated mathematical formulas to which they cannot relate. This chapter takes a different approach.

What Is a Budget?

At the heart of the matter with budgets is a fundamental definition: *a budget is a plan for the receipt and expenditure of funds*. From the perspective of the school leader and policy maker, this simple statement encompasses the totality of the arcane world of budgets in education. The key term in the statement is "plan," which implies that there is some action anticipated by the budget. This is particularly true for budgets in education, which use program budgeting. The most ubiquitous budgeting approach in education is incremental budgeting, which builds on line-item budgets from year to year by modifying the budget based on revenue projections, program priorities and objectives.

This budgeting system ties resources to action to enable some program-based activity. For example, money is designated (*allocated*) in a school district budget (*to a line item*) to support the music instruction program. Behind this dollar amount will be a detailed program for delivering music instruction in the school district, which might include everything from teacher salaries, to purchasing musical instruments, to transportation costs for travel to the state marching band competition. Program budgeting allows for such level of detail and is manageable in even the largest school districts because of incremental budgeting.

As esoteric as individual school district budget practices seem, there is remarkable similarity across the nation among the fifteen thousand school districts in the country. This is due in large part to efforts by the school districts, states, the federal government and the accounting profession to promote uniform budgeting practices and definition of terms. This has greatly facilitated the ability of the federal government, states and school districts to aggregate and analyze financial data for any number of purposes. For example, a more detailed explanation of program budgeting, other budgeting approaches and the terms used in this chapter can be found in the resource *Financial Accounting for State and Local School Systems: 2009 Edition* (Allison, Honegger and Johnson, 2009) and at this website maintained by the National Center for Education Statistics (http://nces.ed.gov/pubsearch/pubsinfo.asp?pubid=2009325).

What Does a Budget Do?

To the surprise of the novice, a budget is much more than a bunch of numbers intended to pay for stuff. A budget, it turns out, is a multidimensional instrument. It has many important functions related to the operation of the education system. It also has many functions as a tool for the management of the education enterprise, i.e., school, district, state PK–12 system or federal education program. Listed below are some key examples of how a budget serves as a tool for school leaders.

Picture 11.1   A good budget system helps to avoid many problems.

Organize

A budget serves as a tool for organizing the information and ideas that make up the intended activity or set of activities. This organizing function applies whether considering the state budget for PK–12 education drafted by the legislature or a school-level budget for instructional supplies. In both cases the budget provides coherence to cycles of getting and spending money. This organizing aspect of a budget also facilitates the many other dimensions of a budget: planning, legal use, monitoring, reporting and auditing.

Planning

The budget tends to bring flights of fancy down to earth. It is not uncommon, and sometimes it is even desirable, to have a free flow of ideas when planning for new or ongoing education programs. Such brainstorming sessions offer the pros-pect of capturing wonderful ideas that might make a big difference in reaching the educational mission. But ultimately, there has to be a way to pay for things and the preparation of a budget causes ideas to be prioritized. The budget distinguishes among the desirable and the essential. In this way it actualizes the program.

Transparency

In public education the budget is a vehicle that fosters the transparency of the en-tity's operation. State and school district budgets are public records and as such are subject to scrutiny by parents, employee groups, the news media and the public. In most states, school districts must adopt or certify their budgets in an open board meeting before submitting them to the next level of review. Copies of the budget are public information available to all. Some communities even publish the entire budget in the local newspaper. This transparency serves to front-load the budget process to ensure the legal and allowable use of money. Furthermore, this transpar-ency is essential to building trust between the community, who foots the bill, and the government agency that spends the money.

Monitoring

Following the theme that a budget is part of a plan for action, it can also serve as a mechanism for monitoring. In this way the budget provides information about the progress of the plan. Consider this example of a school district budget that is estab-lished for the education of new learners of English. The timeframe for the budget is an academic and a *fiscal year* (a twelve-month accounting period) in this example, which starts July 1. The budget shows that 80 percent of the funds are earmarked for personnel costs. Upon reviewing the budget in November, the superintendent observes that no money has been spent from the budget for personnel costs. What might this mean? What should the superintendent do upon making this observation?

The reverse of this example might be if the superintendent observes that rev-enue for the budget is not coming into the district as scheduled. This would clearly be a cause for follow-up by the administrator. So now the personnel expenditures are on schedule, but the superintendent sees that projections for the receipt of revenue for the budget are way behind schedule. Here again is cause for investiga-tion and a potentially serious matter. Teachers have been hired and employment contracts issued. The obligation of the school district to pay the teachers is not reversible. They will be paid even if it must come from some other funding source. Thus, keeping track of, i.e., monitoring, the progress of expenditures and receipts can provide the administrator or policy leader much vital information.

Reporting

Reporting is yet another purpose of a budget. This aspect of a budget serves to keep tabs on the progress of the budget implementation over the course of the budget term, usually a fiscal year. The reporting piece involves the distribution of information on a periodic basis about the status of the budget to all appropriate parties, e.g., the budget administrator, the budget administrator's supervisor, accounting officers and governing officials. A common format for budget reporting during the year the budget is operational is to report receipts and expenditures by line item along with percentages. For example, if a budget from a federal grant was to bring in $1 million, paid out over twelve months, one line in the report would display the percent of revenue received to date.

Another common practice is to distribute reports on a monthly basis. With diligent monitoring, as described in the section above, many program and money problems can be avoided. School boards typically receive a budget report from the superintendent about once a month. The level of detail is usually limited to the several *funds* (a fund is a set of accounts designated for a specific purpose, e.g., transportation, school lunch, general operating) maintained by the school district. More detailed reporting is provided when requested or needed. In a similar fashion, the state, often coming from the state treasurer and/or state education agency, reports on PK–12 school funds. In this way the billions of dollars appropriated for the state education system are tracked to ensure that the budget, i.e., plan for funding the schools, is proceeding as envisioned by the legislature and governor.

Building principals, department chairs and program coordinators also typically get periodic budget reports. Not surprisingly, the purpose is the same—to ensure the program is going as planned and the money to fund the activity is available and being spent as approved.

Auditing

Audits occur at multiple levels for multiple purposes by multiple entities. An *audit* is, simply, an inspection. Budgets are routinely subject to fiscal audits, which focus on the money aspect of a budget. In some cases fiscal audits are combined with program or compliance audits to verify funding was spent for approved activities. School district budgets are subject to *independent audits* annually, i.e., by someone from outside the school district. These are conducted by qualified auditors, usually supervised by or directly carried out by certified public accountants (CPAs), who are licensed professionals. Independent accounting firms are typically hired to conduct such audits, and some states have agencies devoted to this purpose.

Audits are often conducted on behalf of one level of government for another level. For example, the state education agency "audits" enrollment figures reported to it by local school districts. Similarly, when federal funds are involved, the state may audit part of a program on behalf of the federal government. Common practice is for auditors to build on the work of audits conducted at lower levels.

A main focus of the annual independent audit is to uncover fraud, waste or abuse. Independent audits are probably as old as civilization itself. It is easy to imagine an ancient king or Pharaoh having a trusted third party to double-check the figures presented to him by the servant responsible for the kingdom's treasury. The story of Archimedes comes to mind, the ancient Greek mathematician and inventor who proved the new gold crown for the king was not solid gold but had been debased by the smith with silver. Unfortunately, stealing is a predictable occurrence when people and money are involved, and audits serve to uncover theft.

Beyond fraud, audits help to improve the prudent use of funds through the detection of wasteful spending. Just as with stealing, it is also common for people to spend other people's money (i.e., the taxpayer's money) with disregard. While not theft, it is a poor practice that squanders a valuable resource. The paragon example of waste in recent times is the audit of the U.S. Air Force budget by Congress, in which the audit discovered the Air Force was purchasing $700 toilet seats. Waste can occur through poor employee work habits or poor systems within the organization, e.g., not soliciting bids from vendors for big-money purchases. Waste comes in many forms, from inefficient use of resources to extravagant spending.

Abuse is also a concern of fiscal audits. While not quite stealing or waste, abuse is the practice of bending the rules and cutting corners. Abuse happens when an organization or individual exploits a loophole, bends the rules or follows the letter but not the spirit of the law, policy, rule or accepted practice. In education, abuse is often seen in federal education programs, for example, when federal funds supplant local dollars or when employees paid for a specific purpose, say Title I teachers, are used for general education part of the day.

Types of Budgeting Systems

There are many budgeting methods in education that have emerged over the decades: performance budgeting, program planning budgeting (PPB), zero-based budgeting (ZBB), site-based budgeting, outcome-focused budgeting, and line-item budgeting. Each approach has its advantages and disadvantages; thus, it is not unusual to see blends of the approaches in practice.

Legislatures often use *performance budgeting* because of its ability to quantify expenditures to units of activities or services, e.g., cost of a preschool teacher to number of children served. This method facilitates the addition or diminution of funding, and thus service, in a comprehendible way during the legislative process. But this method faces challenges as the units become harder to quantify reliably. Additionally, legislators frequently want to quantify outcomes in dollars per unit that are often controversial and difficult to establish as causal relationships.

Program Planning Budgeting (PPB)

As the name implies, PPB builds a budget from the program level. Its advantage is that it offers a perspective on a scope of the organization that is comprehensible and can be tied to organizational goals, for example, vocational education. Programs are sets of activities designed to achieve a specific purpose. Yet this approach has

Textbox 11.1 Compounding Problems

Compounding Problems

An affluent, medium-sized school district in the Northwest experienced a series of problems because of a budget miscalculation that snowballed into a very large calamity. The adopted budget for the fiscal year overstated revenue by over $2 million. This equaled about 3.5% of the district's operating budget. The problem came to light in January when preliminary results of the annual independent audit were reported to the superintendent and chief financial officer of the school district. These officials decided to suppress this information pending further investigation. The final audit report was presented to the school board in March with a note that the revenue picture was still being determined.

The district enrollment had been growing fast in recent years, which conditioned the board and superintendent to expect increasing revenues each year. The district also had an aggressive teacher and staff salary schedule that included substantial raises each year to help compete for new employees. But this year several other issues added to the problem: state support to the school district was only increased by 1%, enrollment growth was below projections,; and the district had just closed negotiations on employee contracts that included hefty raises. In essence, the district had overcommitted in employee salaries, saw a decline in anticipated revenue and had a $2 million hole in the budget for which it could not account. The magnitude of the problem was now $6 million or almost 10% of the annual operating budget.

The crisis came to light when a reporter from the local newspaper broke the story. The school board was understandably angry about not being informed by the superintendent. Soon the dominoes started falling: the superintendent fired the chief financial officer and the board fired the superintendent. Several board members lost their seats in the next election and a school bond election to build new schools for the growing district was soundly defeated. But none of this helped solve the school district's budget problems.

Eventually, the new superintendent and board were able to negotiate a loan from the state to help it meet obligations until the district could work its way out of the budget crisis. But the terms of the help were harsh: salary cutbacks for all administrators; employee layoffs, with no raises for teachers or staff; heavy state oversight through an appointed financial committee; and drastic cuts to educational programs. These conditions lasted several years. Additionally, because of a lack of trust on the part of the community, the district was not able to pass a bond, so no new needed schools were built, and it was not able to get approval for a mill levy increase from the voters to help with the operating budget. It took many years for the community to regain trust in the management and governance of the school district.

Everyone suffered because of the budget miscalculation and compounding problems: students, parents, the community, teachers, staff, the school board and administrators.

limitations in its ability to offer discrete information for evaluation when looking at expenditures below the program level, i.e., teacher and student. Program planning budgeting is common in public education and government.

Zero-Based Budgeting (ZBB)

Zero-based budgeting requires the justification of all programs and activities on an annual basis. The advantage of this method is that it forces a review of existing or new programs within the context of the organizational mission. A ranking process is often involved to assist in decision making. In theory ZBB will help phase out unproductive services and programs and free up resources for more important priorities. In practical terms, many state and school district programs cannot be eliminated, so the ZBB exercise can quickly degenerate into a perfunctory ritual. For example, does one really need to justify funding high school English each year?

Site-Based Budgeting

This gained in popularity in recent decades. The concept with this methodology is that budget decisions are best made where the need is best understood, i.e., the school level. Budgets are thus built up from the school (or unit) level. Broad discretion is assumed at the site level and it is often the case that collegial decision making is part of the budget process. However, limited resources and legal and policy constraints often limit site-based discretion. In addition, site-based leaders can become distracted from other priorities, e.g., instructional leadership, because of the demands of managing a detailed budget. Also, when budget crises hit, the tendency is to centralize control of all school district funds in order to make sweeping budget cuts, for example, eliminate some programs or services across the school district.

In contrast to site-based budgeting, *outcome-focused budgeting* has more of an accountability tone. This model looks to results as a means of making budget decisions. On the upside this process can weed out programs that do not advance organizational goals. However, the reverse might also be true, since many programs in education are targeted toward very challenging human circumstances where results are difficult to achieve.

Line-Item Budgeting

Line-item budgeting is the oldest and most ubiquitous budgeting method found among school districts. It is a straightforward approach that relies on historical spending patterns to make decisions about future spending. It also enhances central control of the budget, which can help when quick budget decisions are needed. Its simplicity derives from the incremental nature of the budgeting process. So, for example, employee raises can be worked into the budget in consideration of known historical expenditures for salaries and anticipated revenue increases. In line-item budgeting, forecasting for revenue and expenditures, a critical piece of budget development, is more reality centered than some of the other methods outlined above. Critics of this method, however, point out that it supports a "business as usual" orientation that undermines education reform.

What Is the Budget Process?

Budgets tend to follow a periodic cycle, usually over the course of a fiscal year. Within this cycle are seven interrelated stages (planning, forecasting, allocating, submittal, implementation, review and evaluation), which typically appear around the same time during each fiscal year. In some cases the timing of the stages is specified by law or policy. So, like the spring, summer, autumn and winter seasons of the annual calendar, budgets have changing phases, too. States, intermediate education agencies and school districts operate from a budget calendar, which specifies dates and deadlines for various phases and activities of the budget process. Such calendars will come from state-level agencies like the department of education or state treasurer. When federal funds are involved, the U.S. Department of Education or General Accountability Office may set the calendar.

Preparation

Budgets were defined, above, as a plan for the receipt and expenditure of funds. Thus, the first stage of any budget is the planning stage. Whether it's the legislature considering the state budget for pre-collegiate education or a building principal preparing a school-based budget, planning is an essential first step. Remember that education budgets tend to be mostly program related and as such are designed to support some action. Calculating the match between the desired action and the financial resources needed is necessary to a viable plan. Poor planning will lead to underfunding projects, programs or activities. In contrast, poor planning can also result in wasting resources by overbudgeting, thus forgoing or undercutting other important educational programs.

Strategic planning, whether at the state or school district level, is a valuable tool for identifying and actualizing a viable organizational mission and critical goals. A budget that articulates from a strategic plan has a better chance of meeting desired outcomes. Annual budget planning should occur within the context of revisiting the organization's or system's adopted strategic plan. The practical aspects of budget preparation present pressures that are often unrelated to organizational goals and priorities. These pressures tend to be of a more personal nature, like jobs, contracts and positions for specific individuals. Without a well-designed strategic plan, budget planning can degenerate into a series of base political trade-offs.

Forecasting

From the state to the school district level, forecasting revenue receipts is a central part of the budget planning process. In public education, this often involves projecting revenues to be received in a subsequent fiscal year. For example, school districts submit their budget to the state well in advance of knowing several key pieces of information, i.e., the precise amount of state aid to be received, the precise amount of local tax revenue to be generated, the precise number of students enrolling next academic year.

In addition, it is not uncommon for school districts in states that allow public employees to bargain collectively to have unsettled employee contracts at the time of completing their budget, thus not knowing what their salary obligations will be next year (see Textbox 11.2). Some states require the school district to

---

Textbox 11.2 It Cannot Be So!

It Cannot Be So!

A group of corporate chief executive officers (CEOs), many of whom were from Fortune 100 companies was gathered in a major Midwest metropolitan center to participate in their state's school business partnership (SBP). This group had been organized at the behest of the state legislature as a result of pressure from the business community about the condition of education in the state. One of the major issues of the corporate leaders was the amount of money being spent on PK–12 education. They were skeptical of the needs expressed and the veracity of the numbers. With regard to budgets, they thought much of the money problems of the education system could be resolved through greater efficiency.

One of the first orders of business was a presentation by the state superintendent, i.e., chief state school officer. The topic of the presentation was the budget process used to build the State Board of Education's budget request to the legislature. The centerpiece of this budget request was the tens of billions of dollars requested for state aid to school districts. As the state superintendent explained the budget process to the CEOs, one could see expressions of doubt and frustration appear on their faces.

At the conclusion of the presentation, one CEO raised his hand to challenge what he had heard. "So let me get this straight," he intoned, "you expect us to believe that you submit a budget request to the legislature on behalf of all the school districts in the state, yet you don't know how many students will be enrolled next year, what salary raises will be, what ending balances districts will have from their prior year's budgets, what local revenue will be generated based on the value of local property tax or how much the state will collect in state taxes?" The state superintendent simply replied, "Yes, you are correct."

There was a period of silence after the response as many of these corporate leaders tried to mentally process what they had heard. Many heads were tilted and mouths agape. Leaders of some of the biggest corporations in the world were stunned at the complexity and challenges of the PK–12 budget process. "How can you plan in such a chaotic environment?" they wondered. That day these business leaders had a newfound respect for education policy leaders and school administrators.

---

submit its budget to a vote of the public as a means of getting community support for the tax to be levied locally. A "no" vote by the public requires a return to the planning process.

Allocation

Once revenues are known or estimated, the plan of activities and projects can take place in detail. This part of planning *allocates* (distributes among the line items) dollar amounts for various actions, e.g., salaries, materials, equipment. Allocating expected funding within the budget is an act of prioritizing among competing programs and activities in an environment of limited resources. Careful planning makes for viable programs and prudent use of funds.

Submittal

The next step in the budget process is to present the budget for review and approval. This review and approval can range from the very formal to the routine. Submittal commonly requires one level of the system to put forward a budget to a higher level of authority. Some examples are: a department chairperson in a school gives the principal a budget request; a school principal submits a budget to the superintendent; the superintendent submits the district budget to the school board; the school district submits its budget to the state; the state education agency presents a budget to the state legislature; a state submits a budget along with its state plan to the federal government for grant funding.

Submitted budgets are scrutinized for accuracy and appropriateness. This is almost like an audit that takes place before the money is spent. The scope of the review generally focuses on the accuracy of revenue projections, the allowability of proposed expenditures and any obvious instances for potential fraud, waste or abuse.

Approval

Part two of the submittal process is approval. When the higher authority approves the submitted budget, it is sanctioning the proposed expenditures, and by extension the activities they fund. In formal settings, the approving authority may be fulfilling some statutory obligation or policy requirement. In less formal settings, there won't be public hearings or voting to approve the budget, but typically there is a look at the budget and an acknowledgement that the submitter can go ahead and spend.

Implementation

Once a budget is approved, then education leaders execute the budget by spending money on the planned and approved programs, activities, material, equipment, facilities or personnel outlined in the budget. This is the follow-through phase. Now the budget enables action.

Review

As the budget is being implemented, it is important to maintain oversight of receipts and expenditures. The review phase can, and sometimes should, be both ongoing and periodic. Ongoing reviews occur as part of internal controls, e.g., a required signature by a designated administrator before a major purchase. In this example, the administrators look over the budget to ensure there is sufficient revenue to cover the purchase and the expenditure is appropriate to the budget.

Formal reviews can take place in the form of monthly budget reports to budget managers or in budget meetings. In the latter case, this might be the school board going over the district budget with the superintendent and school district financial officer, a legislative subcommittee getting a presentation by their staff or the state education agency, or a principal meeting with department chairs.

The review process serves to keep the plan behind the budget on track and avoid financial problems before they occur or get out of control. If problems are detected, adjustments can be made before it is too late. Perhaps revenue projections were too optimistic, anticipated costs too high or quantities of needed materials and equipment miscalculated. The review process affords time for adjustments to the budget. Depending on the magnitude of the adjustments and the parameters of the approval, changes to a budget may require resubmittal. In such cases, the review and approval process is repeated.

Evaluation

Wise individuals with budget responsibility will take time at the end of a budget cycle to evaluate how well the budget did. Was the plan on target? Were forecasts and projections accurate? Were funds spent efficiently and effectively?

The audit process, described in another section of this chapter, should also serve as part of the evaluation. In more extensive, formal independent audits, it is common to get a "management letter" as part of the budget, which outlines strengths and weaknesses in budget procedures or other fiscal matters. This information is critical to improvement and forestalling problems.

What should be apparent from all this is that budgeting in public education is subject to ongoing scrutiny. The checks and balances of the budget system are apparent from the statehouse to the schoolhouse. Similarly, good practice requires that school districts also adopt equivalent budgeting policies and practices to achieve the same ends of accountability and effectiveness.

What Are the Parts of a Budget?

School budgets exist within a hierarchical system. This system was designed and is maintained by accounting professionals, government officials and educational leaders. The body that oversees this system is the Governmental Accounting Standards Board (GASB). This group has created the school accounting system within the framework of the Standards for Government Accounting, which is in the still broader professional standards known as generally accepted accounting principles (GAAP). The work of this group is facilitated by the National Center for Education Statistics, which has compiled these standards, systems, procedures and guidance into a publication called *Financial Accounting for State and Local School Systems* (2009). Most school personnel who work with budgets on a daily basis refer to it simply as *The Handbook*, which you are encouraged to look over. A more detailed explanation for many of the concepts, terms and definitions presented in this chapter can be found in *The Handbook*. Additionally, individual states will have their unique budgeting requirements that mesh with these standards.

Another important document is the *chart of accounts*. This provides a detailed guide to organizing the array of categories and subcategories used in school budgeting. With the chart of accounts, all revenue can be placed in its proper fund and tracked through the system. Each individual expenditure will have a unique identity and can be explained within this system.

A third resource document is the *budget calendar*. This document establishes the budget cycle. The fiscal year is designated here, along with major budgetary milestones like reports and audits. Due dates are specified and announced. This way, a school district will know when expenditures can take place within a budget, when encumbrances may occur and when obligations must be reconciled. The budget calendar also helps to bring order to the system.

These tools—*The Handbook,* chart of accounts and budget calendar—help policy leaders and education managers keep track of the billions of dollars spent annually on our education system. The budget system put forward by GASB follows a rational approach whose aim is to help education policy leaders and school leaders by:

o Providing a system of classification in order to extract meaningful fiscal information that is comprehensive and uniform;
o Helping state and district leaders comply with GAAP;
o Facilitating reporting to the public, the media and various levels of government, thus supporting accountability throughout the education system.

The anatomy of an education budget highlights the accountability aspect of school budgets, which serves to make clear the financial condition of the state or school district; compliance with laws, rules and policy; and the prudent use of public money.

At the heart of the school budget are funds. A *fund* is a set of accounts designated for a specific purpose, like a transportation fund. School districts typically have major fund types under which lesser funds are established.

Major fund types within the typical school district budget are: governmental funds, proprietary funds and fiduciary funds. The *governmental fund types* consist of the general fund, special revenue fund, debt service fund, capital projects fund and permanent fund. As defined above, these funds have a specific purpose. The general fund is the operating budget for the school district. This is where the light bill is paid, teachers' salaries are recorded and pencils are purchased. The special revenue fund is the place that grants money—for example, from the state or federal government—is received. The debt service fund, as the name implies, is where taxes are collected to pay back (or service) school district debt, for example, when the district sells bonds to pay for the construction of a new school. The capital projects fund is used when constructing facilities. The permanent fund helps to account for restricted money from which only earned interest can be spent.

The *proprietary fund types* are where fees charged by the school district are collected and disbursed. This would be done through an enterprise fund, designed to receive and spend gate receipts from athletic events, for example, or money from renting school facilities to outside groups. The *fiduciary fund types* serve as the place for the school district to hold money on behalf of a group. A common example is when a school district maintains a pension program for its employees or a scholarship fund for high school graduates of the school district.

Other funds can be and are established. In most cases, sub-funds are created within major fund types and lesser funds are built within these (see Figure 11.1 for an illustration). An example of this is when a school district chooses to establish a separate fund for each construction project it has underway. In this case the district would have a capital projects fund among its governmental fund types, and within the capital projects fund it might set up a sub-fund for the new high school, a sub-fund for the new elementary school and a sub-fund for the new curriculum center. States often impose fund requirements on local school districts in pursuit of their interest for segregating money for reporting and accounting purposes.

The program is the next level within the budget hierarchy. Programs exist within funds. A *program* is a set of activities or procedures focused on an objective

❖ **Major Fund Type**

  ➢ **Governmental**

    ▪ **Capital Projects**

      • **New High School Construction Fund**

      • **New Elementary School Construction Fund**

      • **New Curriculum Center Construction Fund**

Figure 11.1   Example of the hierarchy of funds.

or objectives. The budget structure has nine program areas: regular education, special education, vocational education, other instructional (PK–12), non-public school, adult and continuing education, community/junior college, community services and co-curricular/extracurricular activities.

The program structure allows states and school districts to plan for and report expenditures in a manner that helps budget analysis and future planning. As an alternative to or in conjunction with the program classification, states and school districts use what is called a *function/object* approach. The *function* classification is defined as activities for which services or materials are purchased. The function categories include: instruction, support services, operation of non-instructional areas, facilities acquisition and construction, and debt service. Each function can have many sub-functions.

Subordinate to each program and function is an *object*, which is used to describe the planned or actual expenditure for a service or commodity. Objects are divided into nine major categories: personnel services-salaries; personnel services-benefits; purchased professional and technical services; purchased property services; other purchased services; supplies; property; debts service and miscellaneous; and other items. These objects each have many sub-objects, which are specified in *The Handbook* to provide extensive guidance to states and schools. In this way financial data can be collected across a state and across the country with a fair degree of accuracy. This is how reports on teacher salaries are compiled or information on the amount of money going to the acquisition of new computers for classroom use is collected.

Funds, programs, functions and objects are organized through a numbered coding system. In this manner a particular purchased service or item can be budgeted to a particular area or charged to the appropriate budget. A common reaction to seeing a school district budget with all its coding numbers and dollar allocations is to be overwhelmed; the eyeballs roll back or start spinning like a Las Vegas slot machine. But with a few minutes of explanation, policy and school leaders can attain a working knowledge of any of these budgets.

An easy way to relate to the budget coding is to think of a numbering system with which you are already very familiar—the telephone system. Say for example you are traveling through Europe and you wish to call home to Seattle, Washington. To make the call from your hotel room to your home, you must find a way to bypass the millions of other possible telephones you could be connected to in the world. This is done by giving each telephone a unique number. The call follows a set pattern: country code, area code, three-digit exchange number and four-digit telephone number. In this case it would look something like this: 01-206-555-3450, a very familiar pattern.

Budget coding accomplishes the same thing. Consider the example of a salary for a third-grade teacher. The level of detail can run from the fund level to the individual school level if desired. In this example it will be kept simpler: 01-100-1000-100. You may already recognize the coding: 01 is the general fund, 100 is regular elementary/secondary programs, 1000 is instruction in the function code and 100 is personnel services-salaries in the object code. Because the program, function and object codes are comprised of several digits, states and school districts

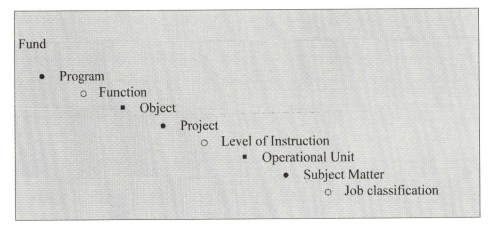

Fund

- Program
  o Function
    ▪ Object
      • Project
        o Level of Instruction
          ▪ Operational Unit
            • Subject Matter
              o Job classification

Figure 11.2   Expenditure account code hierarchy.

can use the available space to further delineate the coding. So in the example for an elementary teacher's salary, the district could embellish the coding by designating a program code for elementary as 1100, middle school as 1200, and high school as 1300, etc. Even individual schools can be worked into the coding system for the school district if they wish (see Figure 11.2).

Revenue

Putting together a budget is merely an exercise unless there is a revenue stream. The revenue fuels the budget and makes the planned actions possible. School districts can have many revenue sources; however, the most common and significant for the General Fund (i.e., operating budget) are state sources and local sources. For some school districts—for example, those with a lot of federal activity like military bases or Indian reservations—federal sources can also be a significant source of revenue for the general fund.

As was noted in the section on funds, the special revenue fund is where school districts usually recognize revenue from federal grant programs such as Title I or special education. A separate fund helps the school district keep federal money segregated from revenue from other sources. This is important since most of the federal grant programs require the school district, and the state for that matter, to account for the grant funds without ambiguity. Thus, commingling federal grant money with local and state money is prohibited in most cases.

As with planned expenditures, revenues are assigned specific line items in the budget. Revenues also have designated coding to identify the source of the money, e.g., 1110 is used for *ad valorem taxes* (taxes on assessed value of real and personal property within the school district); 1120 is for sales and use tax collected locally; 3100 is for unrestricted grant-in-aid from the state, such as state equalization funding.

Each fund will have its own source(s) of revenue and as such will have a unique code for the revenue. In this way revenues are easily tracked to the appropriate budget, and accounting for the proper use of the money is readily available. As with the telephone analogy, the correct revenue code ensures the money goes to the correct budget, and thus is spent for the correct purpose. Here is one more example to illustrate the point. When a school district sells bonds to finance the construction of a school building, it will code the revenue as 5110 in the debt service fund, in the amount of bonds sold. However, this money must be transferred to the capital projects fund in order to pay construction costs. The revenue code 5200 is used in the capital projects fund to show the transfer in from the bond fund. Each of these steps makes it easy to track the movement of the money from each source and reconstruct the movement after the fact, for example, during an audit.

Managing a Budget

Education policy leaders and school administrators each have various levels of responsibility with regard to the money used for public education. As such the level of financial management knowledge required of each individual varies. Clearly, the chief financial officer of a school district should have a high level of expertise. By contrast a member of the local board of education should have a level of understanding appropriate to his or her oversight role.

It is not within the scope of this textbook to delineate the array of roles and responsibilities for budget management for all the responsible parties within the education system. But listed here are some broad guidelines that should be used by everyone from the statehouse to the schoolhouse in managing education dollars.

Provide Training

Whether a legislator, state board member, local trustee, superintendent, school principal or department chair, any individual with budget responsibility should be trained to the extent appropriate for his or her role. Each higher level in the system has an obligation to see that training is available to those with responsibility for public funds within their organization and in the organization it oversees. The training should be routinely scheduled, ongoing and open to anyone with an interest.

Have Clear Fiscal Policies

There is an abundance of laws, regulations, policies and guidance from governing authorities and the accounting profession; these should be the basis of budgetary policy within the organization. This information should be learned and followed.

Promote Transparency

Public funds should never be managed in secret. Budget information that is accessible to all with an interest or curiosity is less likely to be subject to fraud, waste or abuse. Secretiveness surrounding a budget should raise suspicion among those with budgetary oversight or responsibility.

Control Access

This guideline pertains to budget authority, i.e., only a limited number of personnel should be able to authorize expenditures or transfer funding within a budget.

Require Multiple Sign-Offs

A degree of redundancy with regard to review and approval of expenditures will help avoid problems like theft or misappropriations.

Maintain a Paper Trail

All receipts and expenditures should be easily traceable in and out of the budget. This requires a coherent and viable system of recording for every budget, and each budget transaction, whether the state education fund or an elementary school milk money fund. Checks, warrants, vouchers, receipts and other paper (or electronic) documentation are essential.

Monitor Budget Development and Implementation

It is better to catch problems while they are small. Routine budget reviews should be a part of the organization's operations, and each level of the system should be involved as appropriate.

Avoid Cash

Cash is a magnet for problems. When collected it should be recorded and banked as soon as possible. Storing cash on school premises is an invitation to abuse, outright theft or worse. It is not uncommon for a thief to destroy thousands of dollars of school property to get at a few dollars of cash. Having cash around is a bad practice that often invites abuse by employees, such as "borrowing."

Use Proper Accounting Methods

Common standard accounting practices require school districts to use a method of accounting called *modified accrual*. This approach calls for expenditures to be recorded in the budget—as soon as the obligation is made—and revenues to be recognized only upon receipt. The purpose of this approach is to ensure only resources in hand are spent and overspending of the budget is avoided.

Therefore, an entire teacher's salary will be shown as "encumbered" in the budget from the beginning of the fiscal year, even though it is only October and the teacher has been paid two-ninths of his salary. Similarly, the budget may show a deficiency because anticipated receipts (revenue) are being received over the course of the year and not all at once.

Picture 11.2   Avoid having cash around—bank it every day.

Cash accounting is sometimes allowed, but should be used only in limited circumstances. Typically these would be budgets with few transactions for revenue and expenditures. Cash accounting does not typically record obligations or encumbrances, but merely shows money in and money out. The danger of overcommitting the budget is much higher in a cash accounting system because projecting revenue and expenditures can be a problem.

Spend Wisely
Each organization and each level in the organization should be committed to spending taxpayer money prudently. Part of being a policy leader in education is assuming the role of trustee of a public resource. Being an education leader means using the resources given to your organization in the most effective way possible.

Summary
The purpose of this chapter was to demystify education budgets and the budgeting process. While you will not earn your accountant's green eyeshade and sleeve garters based on this information, you should have a better grasp of the system as a whole, and the tools available to those who allocate resources for schools and those who spend those resources. School budgeting exists within a comprehensive system that brings order to complex and extensive funding and expenditure activities. It is

designed to promote the prudent use of the public's money and avoid fraud, waste and abuse.

With a little time and patience school leaders and policy makers can master school budgets and become comfortable with the budget process. There are numerous resources and tools available to help in understanding school finance and budgets. These resources are easy to access and use.

In public education, funding enables programs. Programs are designed to accomplish some educational or support goal. Education budgets are commonly organized around programs.

References

Allison, G. S., Honegger, S. D., and Johnson, F. (2009). *Financial accounting for state and local school systems: 2009 edition* (NCES 2009-325). Washington, DC: National Center for Education Statistics, Institute of Education Sciences, U.S. Department of Education. Retrieved from http://nces.ed.gov/pubs2004/h2r2/ch_3.asp retrieved 12/10/2009.

Dickens, C. (1843). *A Christmas carol in prose, being a ghost story of Christmas.* London: Chapman and Hall.

School Facilities

12

Aim of the Chapter

IN THIS CHAPTER THE CONCEPTS OF PLANNING, designing, constructing, remodeling, maintaining and financing school facilities are examined from the perspectives of the school leader and policy maker. The important relationship between educational programs and school buildings is explored. Cost-benefit analysis is considered within the context of facilities planning and utilization. The aim of the chapter is to underscore and understand the scope of this aspect of education finance.

Introduction

School facilities represent an enormous investment by the American people in their community's infrastructure. Yet most citizens don't think of their neighborhood schools as big capital investments. Usually, they only think of the cost of school facilities when they are asked by their local school board to vote for a bond levy to pay for expenses related to school construction. That is when the idea that school buildings cost many millions of dollars enters the public's consciousness.

Yet beyond the cost of construction, school buildings reflect the history of a community and often the attitudes of its people toward public education at a given point in time. Typically, a new school is a source of pride in most places. It serves as a reaffirmation of commitment to the community's future and to the children and youth of the area. New schools are integral to the viability of the town or neighborhood as a whole. The result is that in some places schools are built as models of austerity and practicality, while in other communities they are examples of modernity and innovation.

As is the case with many of the chapters in this book, it is hard to serve all the needs of policy leaders, administrators, educators and stakeholders with a single approach to the topics covered. This is due in part to the variety of circumstances

Picture 12.1   Communities take pride in their school facilities.

in which school districts exist: one school or hundreds of schools; rural, suburban, small-town or urban settings; heavy state involvement in school construction or little state oversight; state support for school construction or completely local responsibility. The chapter is written in a manner that captures the general sense of the issues, processes and systems associated with school facilities planning and financing. It must be understood that the specifics for each school and school district will be shaped by local conditions, i.e., state policy and law.

Across the Nation

There are over ninety thousand school buildings throughout the country. Each school sits on a lot, which can vary in size from a fraction of a city block to dozens of acres. Some schools are built vertically with multiple stories, while others sprawl over a large area. One-room schoolhouses can still be found scattered throughout remote rural parts of the United States. Mega-school complexes, which look like university campuses or corporate developments, can be seen in urban and suburban locales.

This great national asset, the public schools, has been accumulated over the entire history of our nation, community by community and generation by generation. The collective value of the land, physical plant and equipment held by the public schools ranges into the trillions of dollars. These community assets represent an inheritance passed down from prior generations to contemporary society. Chapter 2 of this textbook presented the historical view of the development of school finance. Recall that even before the national government was fully formed, the Continental Congress passed the Ordinances of 1784, 1785 and 1787, which specifically provided for schools. Schools are a legacy of our values and aspirations as a people.

As one travels within a community, it is easy to see this history of American education, and the nation's history, in the façades of the schoolhouses. The gothic columns of the old downtown high school speak to the enormous community pride of generations past as they committed to that new idea of secondary education for all youth. Many a plain red brick schoolhouse from the depression era of the twentieth century is still in use today, testament of the Works Progress Administration (WPA) of President Roosevelt's New Deal. The open and modern lines of the 1950s- and 1960s-built schools speak to the unbounded optimism of the new world leader and our hope for the future. Many communities recognize this history and have nurtured this historical legacy by setting aside old schoolhouses for historic preservation.

But schoolhouses, like any physical asset, will deteriorate over time and must be maintained, rehabilitated or replaced. The American Society of Civil Engineers (2012), in its latest comprehensive estimates, projects that $268 billion was needed to address the repairs, replacement and safety issues in America's schools. More recent studies put estimates even higher. This figure does not include the tens of billions of dollars spent annually for land acquisition, new construction, additions, remodeling, fixed equipment and furniture. In 2009 this accounted for over $16.4 billion of expenditures for new construction for PK–12 education (Abramson, 2012).

States and school districts across the country face an array of challenges related to facilities. In areas where there is rapid growth, the need for more classrooms is pressing. School districts like Clark County, Nevada, where the population exploded in the 1980s, 1990s and 2000s, went through almost a decade of opening new schools on a monthly basis, Abramson, 2012. Douglas County, Colorado, led the nation as the fastest-growing school district at the turn of the century as it moved from a sleepy rural county to a part of the Denver metropolitan area.

On the other hand, school districts with declining enrollments face the mounting costs associated with excess capacity, which leads to inefficiencies in expenditures for facilities maintenance and repair. Underutilized buildings are a drain on school district budgets, and the obvious, simple answer—shut down some buildings—is one of the toughest political challenges school boards and superintendents can face. Declining enrollments are often a bigger challenge than rapid growth. They tend to be precipitous, difficult to gauge and associated with economic decline. Furthermore, the symbolism of a boarded up school can be the ultimate image of defeat for a town or neighborhood.

Unlike operating budgets, money to build schools or undertake major remodeling projects is not routinely dispensed to school districts. In those states where such funds are distributed on a formula basis by the state, they are almost always insufficient to meet the local need. Such funding usually requires a voter-approved tax increase and a grant application to the state. And while school districts are expected to spend money from their annual operating budget to pay for maintaining their facilities, the pressure to divert funds for other purposes—e.g., salary increases, health insurance, fuel bills and utilities—is great. Thus, common practice across the country is to postpone repairs, delay maintenance and put off building new schools during tight budget periods. All of those delays end up costing the school district more when the facility's needs are so great they have reached a crisis stage.

Political leaders and school administrators are routinely faced with the dilemma of meeting pressing short-term needs while addressing the longer-term demands of facility construction, maintenance and remodeling. The obligation to plan for the future, and to act on those plans, can be difficult when immediate needs are competing for the same scarce resources. Facing voters with a request for new taxes for school construction, while standing for reelection, requires political courage. Taking action on behalf of generations yet to come—for example, by buying land for far off future school sites—requires mature governance and thoughtful policy leadership.

Instruction and Facilities

The relationship between instruction and required school facilities is an ever evolving one. The one-room schoolhouse served the educational needs of early America when small groups of children walked to the school in their area to be instructed by one teacher. The limited age group served, and limited curriculum, made the one-room school an economical and viable design. Today, pre-collegiate education has developed into a more expansive undertaking with broad curricular offerings and age spans that range from toddlers to adults. Specializations of program offerings and perceived efficiencies in organization (for example, grade-span groupings) have led to the design of facilities in use today. The aphorism by the famous early twentieth-century architect Louis Sullivan that "form follows function" certainly applies to educational facilities (Brainy Quote, 2012). Below are examples of how this applies in schools today.

General Education

The idea that "general education" even exists in contemporary schools is certainly open to debate. General education has come to mean academic instruction for a group of students in a classroom: the basic four walls, a writing board and rows of student desks. But this notion of the general classroom is dated. Small-group instruction, cooperative learning groups, learning stations, mini-labs and exploration centers are routinely found in general education classrooms today. No doubt in some schools, rows of desks might even be looked upon with concern. The challenge for education leaders is to design today's school with sufficient flexibility to meet educational program needs, which are often changing and hard to predict.

Even when a consensus is reached about the basic design of a general classroom, many other design elements still come into play. How does the curriculum change as students move through the grades? What are the design implications of these curricular changes? At what point does a classroom's design exceed its capacity to be flexible in accommodating the curriculum? And finally, before the concrete is poured, implicitly or explicitly, policy makers and school leaders must answer this question: On average, how many students will be served in these general education classrooms?

Grade Levels

Space recommendations for general classrooms vary by grade level. School architects and design personnel will suggest 800 to 1,200 square feet, depending on grade level and anticipated class size. Storage space would be additional. Some rural schools or alternative schools might never expect to see classes of twenty-five or thirty-five students, so their space needs might vary from these norms. For other communities this might be the norm. Primary school classrooms are generally smaller than secondary school classrooms, but here again, the variety of instruction planned for the space might alter this norm. While school planners should consider the recommendations of experts when designing a new structure, in the final analysis, they must design based on the needs, preferences, intended curriculum and financial resources of the individual school.

Professional school architectural associations, state education agencies, building permitting authorities and school construction consultants work from a set of standards or guidelines regarding square footage and expected accouterments for various classroom spaces and grade levels. These represent the contemporary industry standards. They serve as the starting point for school design and should be modified to meet local conditions (California Department of Education, 1997; Florida Department of Education, 2007; Texas Education Agency, 2012; National Clearinghouse for Educational Facilities, 2008).

Special Programs

Curricular programs that are not accommodated in a general classroom require uniquely designed spaces. Such programs include everything from preschool classes to physics laboratories. These learning spaces require particular sizes, configurations and equipment. A small high school gym, for example, can be as little as 4,000 square feet with an additional 2,000 square feet for locker rooms, coaches' offices and storage. Large high schools that serve 3,000, 4,000 or 5,000 students will have multiple gyms that are five times larger. Athletic fields, indoor and outdoor courts, swimming pools and even diving wells can be part of the mix. Classrooms built for special education can vary greatly as well, depending on the population of students to be served, such as severe needs or medically fragile students. Laboratory science classes, occupational education, art studios and technology centers are other

Picture 12.2   The curriculum should dictate the design of the school.

examples of classrooms with distinctive design elements. Music programs, instrumental and vocal, also require particular engineering considerations, like acoustical construction, instrument storage and risers.

Common Areas

The auditorium, library and media centers, cafeteria, teacher preparation rooms, copy centers, administration offices, counseling offices, courtyards and lobbies are examples of common areas. Sometimes these spaces can get short shrift during the planning stage of construction, but as with any poorly planned part of a new school, this is a mistake. Student and faculty needs and expectations should be reflected in the planning of these spaces. Community needs should also be considered when designing common areas. Will voting take place at the school? Will the library be used by the public? Does district policy allow for use of school facilities for community events like flea markets, dances, club meetings, community theater or church services? Will adult education be part of how the school is used?

The implication for school design based on anticipated utilization of the building is vast. For example, an elementary school is typically planned with a small parking lot since it only needs to accommodate faculty and staff. But what happens on parents' night or for an open house? What if the school is used for some other purpose like voting or community education classes? Thus, an elementary school might not be an elementary school after all, but a community facility with multiple uses. This should be reflected in the design of the building.

Changing Architectural Concepts

Travel across the country and visit any school, and you can expect to find certain architectural features common to almost every school. There will be an office, classrooms and a space for students to eat, which may be used for other purposes like a gym or auditorium; perhaps you will find a library and playgrounds or athletic fields. Yet while there are generic aspects to schools, from one to another, there are also unique architectural components to each school that blend the financial realities of the school district with the community's vision of education.

Community pride is a very real and oftentimes significant part of the design of a school. This can have practical implications when looks or the inclusion or exclusion of certain spaces overwhelms available dollars for more practical and needed spaces. For example, it may be that community theater is important in a particular town, and so the design of the new school auditorium may be the focus of a lot of attention. These highly specialized spaces can be very expensive. An auditorium and stage to accommodate a career day speaker versus a Broadway play can be two very different things. A large stage, dressing rooms, orchestra pit, raked seating, lights, rigging, catwalks, sound systems and control rooms can turn a school auditorium into the Shubert Theatre. Such design decisions can easily run into the several millions, if not tens of millions, of dollars. The question then becomes, at what expense? What doesn't get built in order to pay for an elaborate theater?

At the other extreme, the overemphasis on mass education as the purpose of the school will lead to a generic facility that does not consider the unique educational purposes of the school, the future or the community needs. While efficiency is important, it should not be the only driving force in a school's design. Schools should be built with a perspective that telescopes out into the future at least one hundred years. How will the building hold up and how will it serve the community over the generations? Cookie-cutter approaches often turn out to be less efficient over the longer run because building occupants end up modifying the original design at great expense.

Accept that things will change over time. Energy efficiency was not a big consideration for schools built in the 1950s or 1960s. Today it is a major priority along with "being green." The modern "open campus," common in those parts of the country with mild climates, was a great idea at the time it was built. But in the era of Columbine High School-type shootings, access, regress and security are now big concerns when designing a school.

Other tragic incidents remind us of the need to establish and enforce building codes related to school construction and safety. In 1958, ninety-two children and three nuns were lost in a fire at an elementary school in Chicago. The tragedy of the fire at Our Lady of the Angels School led to major reforms and attention to school safety in Illinois and throughout the nation (Huppke, 2008). The 7.9 magnitude earthquake of May 2008 that devastated China revealed the tragic mistake of poor school construction and the disregard of building standards. Estimates of the loss of human life range upwards of seventy thousand people, among whom were ten thousand students. Due to shoddy construction, seven thousand classrooms crumbled to the ground. Because of the timing of the quake, entire school populations were lost in an instant as school buildings collapsed on themselves across the devastated region (*New York Times*, 2009).

Technology in recent decades offers a good example of how fast things can change. As the personal computer became affordable and began to appear in schools across the country, the need for computer labs, more electrical power and the networking of computers became a real design challenge for schools. Hundreds of millions of dollars were spent retrofitting schools to provide computer lab spaces and more electrical outlets, and to string cable for computer networking. Today the personal computer and handheld devices are used very differently from the decades of the 1980s and 1990s. Then, computer laboratories were considered essential, and networking was accomplished by wires. Today, wide-area networks use radio signals to create local networks and link to the World Wide Web.

An architectural design element in a newly constructed school thus represents the best thinking about immediate needs, planned uses of the school and guesses about future trends. No one person, nor one group, alone is likely to have the complete perspective needed to design the best school for the money available. This is why school planning and construction should be a team effort among educators, policy makers, design and construction professionals, and the community.

Facilities Planning

When a school district decides there is a need to build a new school, or a series of new schools, it is setting out on a process that has long-term implications for students, parents, faculty, staff and the community at large. It involves deciding where to build the school, which will affect property values and possibly housing patterns. It involves a long-term financial commitment, which often forecloses other long-term financial commitments. It sets in motion actions that will touch generations to come. The decision to build is one of the critical judgments made by school policy leaders.

School District Strategic Plan

Each school district should be guided in its big and small decisions by a strategic plan developed with broad input from all stakeholders and a thoughtful, studied process. This plan should be viable in that it serves as the touchstone for resource allocations and the purposeful actions of all associated with the school district. The strategic plan should be updated on a regular basis and reflect the beliefs and commitment of the school district policy leaders. The strategic plan expresses the clear direction and priorities of the school district. All decisions, big and small, flow from the strategic plan. Therefore, as the school district approaches the issue of need for school facilities, the path toward decision making and the context for the decision are clear. Planning for new schools occurs within the school district's strategic plan.

When and How to Determine Need

In some ways school districts are like people: they are mostly the same as a classification, yet very different as individuals. Some school districts will come to the conclusion that a new facility is needed based on simple observation: the building is overcrowded, and we have more students than available teaching stations can accommodate—perhaps it's time to think about adding a new school to the district. In another setting a school district will use sophisticated central planning, where demographic and economic trends are tracked and new housing developments are monitored closely. Regardless of the approach, making the right decision at the right time is critical. This is because it is a decision that involves a major investment and long-term consequences. Rarely does a school district get to "do it over" when a mistake is made about new school facilities. And all mistakes in school construction have a built-in punishment: they are very expensive to fix.

Remodel or Build New?

A common dilemma faced by a school district as overcrowding becomes apparent is whether to remodel, add on to an existing building or initiate new construction. A combination of factors must be considered. At the heart of the matter is the cost-benefit analysis, where the school district leadership must determine the best use of available financial resources given the short- and long-term district needs. Additionally, access to hard data about current and future trends is essential to better decision making. While no one can control the future, there are many ways for school district leaders to understand where the district is headed so they can prepare for the future.

The School Survey

One of the most powerful tools available to policy makers, school leaders and stakeholders venturing into the school construction arena is the school survey (Ramirez, 1987; Castaldi, 1994). Developed by school facility specialists in the early part of the twentieth century, the school survey takes a comprehensive look at the school district. It draws on analytical data to shape a broad and deep view of the school district.

The school survey will typically have data related to: community and student demographic trends; enrollment trends and projections; housing patterns and planned development; extant school district facilities inventory, including safety and capacity issues; school district land holdings; current, and planned or desired educational offerings; financial capacity of the school district; school construction costs and trends in the region; cost-benefit analysis related to remodeling and new construction; resource people and organizations available to facilitate decision making; and anything else that might help the leaders and community make a decision about when, if and where to build.

Planning for a New Building

One issue that regularly recycles through the education policy debate arena is the question of school size. Much research and opinion has been written about the question of ideal school size (Ramirez, 1990; Ramirez, 1992). Ultimately, practical decisions must be made about the size of the school to be built. Consideration must be given to enrollment trends, community demographics, housing patterns, curriculum offerings, future school construction plans, cost of construction and district financial resources, to name just a few important factors. In the final analysis it is the policy makers, the local board and administration who should decide what size school is best for the community.

Educators and school board members need not take on the analysis and planning on their own. There are many organizations and professionals who are available to help. These people can provide critical information, which can feed into planning and decision making. Here again, some states may have such personnel within the system of government. In other cases the school district must go out and find this expertise.

Architects who specialize in school design are readily available to help with cost estimates and design recommendations. Often these professionals are willing to help a school district identify some rough figures and preliminary design sketches for free, in the hope of gaining a contract once the decision is made to build. Large school districts with extensive building programs may have "in-house" architects to do this work. In some states, county, regional or state agencies provide this service.

In school districts without in-house capability, the facilities and school survey can be contracted out to a university professor or consultant who specializes in such services. They are typically adept at research, writing and group facilitation, which are essential to a successful school survey report. The state administrators association or school business officials may also offer this service or have referral assistance.

Legal and Financial Considerations

Determining the legal ins and outs of financing a school construction project is generally the purview of a bond counsel. These attorneys are experts at understanding and guiding the school district with regard to local, state and federal legal requirements concerning the debt and financing issues associated with the building of schools. In tandem with the bond counsel, the bond consultant is the expert who will help the district issue debt, usually in the form of bonds, to finance the project. The bond consultant will help determine the parameters of debt that can be taken on by the school district and project the cost of the debt given current and anticipated market conditions. These individuals are important to the process, because a school district does not want to issue debt in an illegal manner or miscalculate how much it can afford to build and how much it will be required to repay. Bond counsels and bond consultants work for a fee taken as a percentage of the amount of debt issued.

Another person with critical information is the building contractor. This is the person who heads the firm that will actually build the school or schools. Generally these people engage with the school district through a competitive bidding process, through which the school district is obligated to select the "lowest responsible bid." Many construction companies specialize in schools and have great expertise regarding design, standards and costs. The building contractor comes on the scene after the decision to build has been made and financing is in place. This individual will provide the detail regarding the cost of the building project, down to the last nail and shingle. Some school districts may choose to hire a construction consultant to help with the construction bid preparation and consultation during the bid review process. Here, again, big school districts with ongoing construction projects may find it more desirable to have these services in-house.

Two very important groups to involve in the facilities planning process are educators and community groups. Educator involvement is critical to ensuring that the school is planned with an instructional focus in mind. Educators are the experts with regard to the relationships among instructional spaces, student behavior and learning.

Parent and community groups are essential so that community values and standards can be reflected in the design of the school or schools. Their input and oversight will build credibility within the broader community. Such groups will also serve as a key communications vehicle to the broader community regarding the design, planning, financing and construction process. Parent and community involvement supports an open process, which is important, particularly when the inevitable snags and unforeseen problems emerge.

Building the New School

A new school will take a year to 18 months to build from the time the decision is made to build to the time the children are seated in classrooms. However, school districts that are building multiple facilities over extended years may often refine the process to get faster building cycles. Many, many factors affect this timeline. Leaders should be very careful about forcing too tight a schedule; thus, good advanced planning is important. The last thing you want is to delay school while the board and superintendent are begging the local building inspector for a certificate of occupancy—or worse, try to hold school on a construction site.

Oversight

It is important that a school district employee, often an assistant superintendent, be assigned general oversight of the project. Unlike a construction superintendent or consultant, this person should be an educator who can make judgments about the 1,001 little decisions that are being made as the school is built. In this way the education perspective is always maintained. Common planning tools like Program Evaluation and Review Technique (PERT) charts and related computer software will help keep track of the project's developments. Such tools help to organize the overall effort and sequence, schedule and time of the project. They are good indicators of whether the project is on course or headed for problems.

Expect Problems

Changes and problems with construction are to be expected. Contractors are used to such issues and are usually willing to negotiate changes or unforeseen problems not covered in the bid contract. It is important here for the architect to sign off on any proposed changes to the design of the school. But keep in mind, poor planning on the part of the school district that results in major or constant changes in the building plan will cost the school district substantially. A good plan is essential to a timely, cost-effective and successful project.

Get Ready for School Opening

Anywhere from a year to six months prior to completion, a site administrator should be designated for the new school. This person will undertake the process of making sure the school is properly furnished and equipped. He or she will also start the process of hiring faculty and staff for the new school or transition personnel from the old to the new facility. This site administrator is responsible for ensuring that all the small and large details associated with opening and running the school are accounted for. The goal is to have the new school start smoothly, with minimal confusion and chaos.

Expect More Problems

Construction problems with a new building are inevitable. Common practice is for the builder to be available for a year or more to address problems and needed fixes after construction. This is yet another reason to be careful about selecting a builder: you will be "living" with them for a long time. A builder's professional reputation regarding fairness, quality of work and responsiveness to client concerns is very important. This reputation should be investigated thoroughly before awarding a contract.

Remodeling

For various reasons school districts often face the necessity of remodeling an existing facility. These reasons can range from a retrofit of an existing building to accommodate a new use—for example, converting a middle school to an elementary school—to an emergency situation that requires prompt action. Remodeling is a common occurrence for school districts. However, it is unfortunate when school districts must use this alternative because of a lack of fiscal capacity to meet building needs. Remodeling can be an expensive short-term solution when the real need is new buildings for an expanding enrollment or an obsolete building.

Table 12.1   Recommended square footage for instructional areas.

Type of instructional space	Range in total square feet	Range of square feet per student	Comments
General classroom: primary	850–1,100	42–55	Based on class size of 20 students.
General classroom: secondary	800–1,200	32–48	Based on class size of 25 students.
Music room: chorus	1,250–1,600	50–64	Based on class size of 25.
Music room: instrumental band	1,300–2,000	52–80	Based on class size of 25, exclusive of practice rooms and instrument storage.
Laboratory science	1,200–1,600	48–64	Based on class size of 25; does not include preparation rooms and storage areas.
Gym	3,500–7,000	140–280	Based on a class size of 25, one teaching station, exclusive of storage, locker rooms, showers, bleacher seating, etc.
Library	1,000–?	40–? Should be able to accommodate several classes of students and routine library functions at any one time.	Size depends on size of collection, size of school enrollment and anticipated use, for example, extensive computer stations for students.
Storage	15–50% of gross space		Storage can be accomplished through the use of cabinets, closets or storage rooms and is determined based on need and safety considerations.

Plan Ahead

One way to accommodate growth through remodeling is to build it into new construction. In this case, a school district will build a facility in such a manner that it can be expanded years later when the enrollment catches up. This can be a cost-effective way to handle growth, since mechanical expenses, e.g., heating, cooling, etc., are already in place when the remodeling takes place. But a note of warning is mentioned about this approach: the school site must be sufficiently large to accommodate the ultimate size of the school, so good planning is a must.

Sudden Growth

Growth is frequently handled through the use of temporary spaces, like portable classrooms. These modular classrooms can offer economy and flexibility to the district. They are cheaper than building with brick and mortar and can be moved to new locations as enrollment shifts in the district. The downside of these "temporary" buildings is that they tend to deteriorate and depreciate quickly, can be hard to maintain and are more susceptible to vandalism.

Health and Safety

Emergency remodeling is another very costly expense. Unforeseen health and safety conditions will precipitate the need to remodel. Some examples over the past several decades that have affected schools on a broad scale have been the asbestos abatement mandates and radon gas mitigation programs. Other conditions related to natural disasters will also prompt the government to mandate school remodeling or force school districts to address an emergency need. The rapid rise in the cost of heating and cooling buildings has spurred schools to look anew at conservation retrofits to help save money. The addition of solar and wind power generating capacity is a recent retrofit sweeping the nation.

Disasters

School districts insure their property the way homeowners and businesses do. Hopefully, they are sufficiently covered to replace lost buildings in the aftermath of a disaster. Hurricanes, earthquakes, tornadoes, fires and floods are reported as destroying schools somewhere in the country almost every year. Replacing or re-building under these circumstances is a costly and traumatic undertaking. School leaders have an obligation to make sure the school district is sufficiently insured to recover from a disaster. Insurance policies should be reviewed annually.

Maintenance

Americans have an expectation that their school buildings will be neat, clean and well maintained. The school represents a major community investment and people want to know that their investment is being cared for. Clean and well-maintained buildings also relate to student, staff and faculty health and safety. Medium- and larger-size schools have custodial personnel on staff to handle the need for immedi-ate cleanups during the course of the day. Routine cleaning and non-emergency maintenance is taken care of after school hours. Some school districts will out-source this function to private companies.

Staff Deployment

Recommended staffing in order to keep a building clean and well maintained will vary depending on a number of factors, but square footage is the prime consider-ation. The layout of the building, the materials used for floor and window cov-erings and the geographic location will all come into play. In the Northeast, for example, snow will be a contributing issue for safety and cleanliness, whereas in the Southwest, a constant battle with dust will be the cause. The Association of School Business Officials offers practical information regarding the maintenance of schools (Chan and Richardson, 2005). The National Center for Education Statistics (2008) provides a free downloadable planning guide.

In-House Projects

More extensive maintenance projects require skilled craftsmen to get the job done. It is not unusual for larger school districts to employ a corps of craftsmen like painters, plumbers, glazers and roofers to keep up with facility demands. These specialists not only work on emergency repairs, but also are most often occupied with ongoing maintenance throughout the school district. Some districts will even hire full-time landscapers, gardeners and arborists to see after the grounds.

Cost

The amount of money needed to build a school will vary widely depending on a multitude of factors. While school districts may have broad discretion in whether they want to build a palace or a cabin, some cost elements are beyond anyone's control and subject to market forces. An acre of land in rural South Dakota (without mineral rights) can probably be had for much less than a similar-sized plot in downtown Manhattan. Prevailing wages for construction workers and craftsmen will vary according to the region of the country. The price of materials will fluctuate due to everything from local construction activity to global economic forces. Determining costs for new construction or remodeling is a time-sensitive matter and region-specific. In addition, some school construction projects are subject to the federal Davis-Bacon Act or similarly structured state provisions regarding prevailing wage laws (U.S. Department of Labor, 2012).

The cost to maintain schools is slightly more predictable. On average, school districts across the country spend about 15 percent of revenue for school operations, of which maintenance will be a big part. It is generally held that negligence and deferred maintenance end up costing the district more over time. Deteriorating facilities cost more to bring back to standard and will often end up with costly safety problems. While not a glamorous task, school leaders and policy makers have a responsibility to see that the public trust is upheld by taking proper care of the school district's assets.

Good planning and cost analysis can combine to help a school district get a handle on required expenditures for school construction and maintenance. Local and state school administrator associations are a good source for comparing local construction costs and expenditures for maintenance. Local design and building professionals should also be used when planning for a new school.

Summary

School facilities are an enormous financial investment acquired over generations and handed down as an endowment to contemporary society. Maintaining this investment is an obligation of the current trustees of the schools and school district. Construction costs for new buildings will vary widely depending on planning and local conditions. School maintenance is an essential element of managing a building and school district well. Good planning and the judicious use of design and construction professionals are important to successful leadership in this aspect of school district responsibility.

References

Abramson, P. (2012). New buildings take lion's share of fewer construction dollars. *School Planning and Management*. Retrieved from www.peterli.com/spm/pdfs/SPM-Construction-Report.pdf.

American Society of Civil Engineers (2012). *Report card for America's infrastructure*. Retrieved from http://www.reportcard.org/fact-sheet/schools.

Brainy Quote. (2012). Retrieved from www.brainyquote.com/quotes/authors/l/louis_sullivan.html.

California Department of Education (1997). *Educational specifications: Linking design of school facilities to educational programs*. Sacramento: California Department of Education. Retrieved from cde.ca.gov/ls/fa/sf/documents/edspecs.pdf.

Castaldi, B. (1994). *Educational facilities: Planning, modernization and management*. Boston: Allyn and Bacon.

Chan, T. C., and Richardson, M. D. (2005). *Ins and outs of school facility management: More than bricks and mortar*. Lanham, MD: Rowman and Littlefield Education.

Florida Department of Education (2007). *State requirements for educational facilities*. Tallahassee, FL: Florida Department of Education, Office of Educational Facilities. Retrieved from www.fldoe.org/edfacil/pdf/sref-rule.pdf/

Huppke, R. W. (2008, November 29). *Our Lady of the Angels: The fire the changed everything*. *Chicago Tribune*. Retrieved from http://www.chicagotribune.com/news/local/chi-our-lady-of-the-angels-fire-students-killed,0,6650568.story/

National Center for Education Statistics (2008). *Planning guide for maintaining school facilities*. Retrieved from http://nces.ed.gov/pubs2003/maintenance/index.asp.

National Clearinghouse for Educational Facilities (2008) at the National Institute of Building Sciences 1090 Vermont Ave., NW Suite 700, Washington, D.C. 20005 http://www.ncef.org/.

New York Times (2009, May 6). *Sichuan earthquake*. Retrieved from http://topics.nytimes.com/topics/news/science/topics/earthquakes/sichuan_province_china/index.html.

Ramirez, A. (1987). The educational survey: A powerful resource for school bond elections. *School Business Affairs*, *53*(7), 58–59.

Ramirez, A. (1990). High school size and equality of educational opportunity. *Journal of Rural and Small Schools*, *4*(2), 12–19.

Ramirez, A. (1992). Size, cost and quality of schools and school districts: A matter of context. In Humphrey Institute of Public Affairs (Ed.), *Source book on school and district size, cost, and quality* (p. 72–93). Oak Brook, IL: University of Minnesota, North Central Regional Education Lab.

Texas Education Agency (2008). *Facilities standards*. Austin, TX: Texas Education Agency. Retrieved from www.tea.state.tx.us/index2.aspx?id=5475&menu_id=645.

U.S. Department of Labor (2012). *Davis-Bacon and related acts home page*. Retrieved from http://www.dol.gov/whd/govercontracts/dbra.htm.

Grant Funding

13

Aim of the Chapter

THIS CHAPTER PROVIDES AN OVERVIEW OF GRANT funding programs commonly found in schools. Strategies for writing grant proposals, securing grant funds and managing grants are also provided.

Introduction

Grant funds are a significant part of the financial resources available to education, even though these funds tend to comprise a relatively small percentage of a school district's or school's total available funding. The significance of these funds stems from how schools are able to use the money, which can range from specific targeted purposes to broad school-based discretionary projects.

With few exceptions grant funds are typically supplemental to the regular education program funding and thus can afford an opportunity to a school district or school to support an extra program or service that would not ordinarily be available. Most schools have financial needs that exceed their financial resources. As a result, school administrators are constantly on the alert for grant funding opportunities.

Typology of Grants

Grants come in all types and sizes. They also cover an array of purposes as diverse as the grant funders and recipients themselves. How grants are labeled or classified depends on the characteristics of the grant, for example, how the money can be used or who is eligible to receive it. Classifying a grant can be confusing at times because a grant can have several distinguishing characteristics; thus, it can fall within several classifications at the same time. The entity that receives a grant is called the *grantee, recipient* or *awardee*. The organization that distributes the grant funds is often referred to as the *grantor, funder* or *awarding agency*. Below is an explanation of some of the terms commonly used to describe and classify grants awarded to schools and school districts.

There are two broad groupings of grants:, *formula grants* and *competitive grants*.

Formula Grants

These are funding programs that distribute grant resources to a predetermined recipient according to an established allocation process. The most common formula grant found in schools is Part A of Title I of the Elementary and Secondary Education Act (ESEA), which was reauthorized in 2001 as the No Child Left Behind Act. Over $7 billion of these grant funds are distributed to the states based on a formula included in the enabling legislation created by the U.S. Congress.

The law further specifies how the state is to distribute the money to individual school districts, and then to schools. Formula grants often target specific populations for service, e.g., the disabled or poor. These grants strive to focus on an education-related need, with the expectation that the grant funds will assist or promote a locally funded effort.

Money is commonly allocated based on the number of eligible students in the population. So, in the case of Title I, a state with a higher concentration of children from economically disadvantaged homes will receive proportionately more money than a state and its school districts that are less affected by poverty. Most government grants to pre-collegiate education can be classified as formula grants.

Competitive Grants

These are funding programs that distribute grant resources to targeted recipients based on how well an eligible applicant demonstrates the ability to match some pre-established funding criteria and successfully fulfill the terms of the grant, relative to other applicants. As the name implies, competitive grants presuppose the number of applicants will exceed the number of grants that are awarded; therefore, applicants compete against each other for the grant funds. Some government grants are also awarded on a competitive basis.

Many private sector, eleemosynary and foundation-based grants are competitive grants. Competitive grants frequently have the objective of stimulating new educational practices or services to new populations by providing financial incentives to states, school districts or schools. Therefore, criteria for selecting grant recipients for competitive grants often include a range of items such as: the likelihood of the success of the proposed program; the willingness of the grantee to share program evaluation results or demonstrate program operations; geographic location; available local matching funds; or quality of the staff.

Within the two broad classifications of formula and competitive grants are further subdivisions such as *categorical grants*, *block grants*, *direct grants*, *discretionary grants* and *research grants*.

CATEGORICAL GRANTS   These grants provide financial resources for a particular population, target group or specified purpose. Title I of ESEA and the Individuals with Disabilities Education Act (IDEA), in addition to being formula grants, are categorical programs because their funding is concerned with children from impoverished homes or children with disabilities, respectively. Additionally, because the money from the grant is restricted to a limited number of uses—i.e., instruction in mathematics, reading, writing, staff development, or related services associated with a disabled student's individual education plan—categorical grants typically have a narrow educational purpose. A grant program to assist school libraries with acquisitions to their collection would be an example of a categorical grant. Such grants will explicitly prohibit expenditures for other purposes.

BLOCK GRANTS   Yet another granting approach, block grants can take many forms. What distinguishes the block grant from, say, the categorical grant is that the money can be used for a variety of purposes. Block grants are characterized by the grantee receiving a chunk of funds with broad parameters regarding how the money can be spent.

Sometimes, a legislative body will lump together several categorical programs and give permission to grantees to spend the money at their discretion within any of the purposes of the antecedent categorical programs. While not quite general aid, recipients of block grants usually appreciate the ability to spend money for the broad purposes of the block grant.

DIRECT GRANTS   These grants are made from the grantor agency to a recipient without regard to other potential or similar recipients. These grants can be made because the grantor merely decides to arbitrarily pick a grantee or because of some special circumstance. An example of a direct grant is when a state legislature appropriates money to a school district to build or repair schools as part of a disaster relief effort related to a natural disaster.

DISCRETIONARY GRANTS   This is a term used to define the grantor and that agency's ability to exercise freedom of choice regarding who should receive funding and how the funds should be used. Typically, discretionary grants have a broader purpose, and funds from discretionary grant awards give greater options for spending to fund recipients. Competitive grants and direct grants can sometimes be classified as discretionary grants, when the eligibility criteria and uses for the money are very broad. Discretionary grants are used by the grantor to target some new or innovative program, often on a pilot basis. A recent example is the Race to the Top funding, distributed to states on a competitive basis by the U.S. secretary of education.

RESEARCH GRANTS   As the name implies, these grants have as their objective the discovery of new knowledge. Common examples of research grants are found

Picture 13.1   Grant funds can help solve the resource puzzle for schools.

at universities; for example, a medical school receives a research grant from the National Institutes of Health to investigate smoking habits of a given population as part of a broad national effort in this area. Research grants are frequently also competitive grants and reserved for institutions with highly specialized technical capabilities. Thus, eligibility criteria are very restrictive.

How to Get Grant Funds

There are three principal ways to secure grant funding. The first and most common method is for schools or school districts to be notified by the granting organization that they are eligible recipients for a formula grant or eligible to apply for a competitive grant. This is typical of the relationship between school districts and their state department of education. Such announcements often come as a formal notice in what is called a *request for proposal (RFP)*. The RFP contains all the essential information needed to apply for the grant: who is eligible to apply, how much one can apply for, the expected term of the grant, and by what criteria the grant will be awarded.

A second approach is when the school district or school hunts for grant funding opportunities among potential grantor agencies and organizations. This method might include letters of inquiry, telephone calls or Internet searches. In this situation the school inquires about the possibility of available grants and the school's ability to participate in the program.

A third method for getting grants is to develop a grant program idea directly with a granting agency. A typical scenario in this case is when a school meets with a potential funder, such as a not-for-profit foundation, and suggests an idea for a grant program. Some businesses with community relations functions will respond to this type of contact as well. The key to this method of securing grant money is to understand which organizations have the latitude to award discretionary and direct grants.

Although a more abundant source of funding for schools, government grant proposals tend to require extensive paperwork and other requirements. They are often more complex and difficult to develop. Furthermore, government proposals usually require strict adherence to prescribed formats and can have elaborate reporting requirements for grantees.

Foundations tend to have less elaborate proposal requirements. Grant applications are often shorter, less complicated and easier to complete. Reporting requirements for grantees are correspondingly simpler. Unfortunately, foundations often do not have the large sums of money governments have or limit their funding to smaller target groups.

Corporations and local businesses frequently have the simplest and easiest grant application processes. They also tend to require even less reporting. However, such grantors also tend to offer less money, have a limited scope of grant programs and sometimes retreat from such giving when corporate profits decline.

How Proposals Are Judged

In order to judge the value of a grant program to the school or district, certain requisite organizational management devices should be in place. These organizational management devices include: an organizational vision that helps students, employees and stakeholders maintain a sense of the long-range direction of the district or school; a mission statement that clearly articulates the aim of the organization; and a plan designed to actualize the vision and mission of the district or school.

Under ideal circumstances these devices are developed over time through a collaborative process that involves the policy leaders, employees and stakeholders of the school district or school. These devices work best when they are viable and hold broad-based support. They also serve a valuable and continuous benefit to the organization when they are incorporated into a strategic plan. Here is how these tools apply in the grant solicitation and writing process:

○ Defining your school or school district—Grantors want to know who you are, who you serve and what you do. They also want a sense of where you are headed as an organization. Frequently, grant proposals require such information as part of the formal application process to determine your eligibility to even apply for the grant.

o Determining your need—A good strategic plan helps an organization understand its strengths and weaknesses. This is important during the grant-seeking stage, because it can guide the school to appropriate funders. Additionally, grantors want to be sure their funds will make a difference and not merely serve as an extra resource base for the school district.

o The organization's ability to demonstrate need is often a major criterion for funding decisions. However, it is common for schools to confuse needs and wants (Scriven and Roth, 1978). Understanding the difference between the two is essential to writing good grant proposals and conveying how the grant funds will affect the school. When the teenager tells his parents that he needs a car, the normal reaction from the parent is to question the term "need." Similarly, granting agencies closely scrutinize needs represented in grant proposals.

o Understanding your organization's capacity—A counterbalance to understanding the needs of an organization is to know the capability of a school or school district. The capacity of an organization to fulfill the terms of its grant program is important. A school district must be able to demonstrate it can meet its obligations under the terms of the contract, which is essentially what the grant proposal and grant award become.

o It is a mistake for a school district to overcommit itself in the interest of securing a grant program. It is just as bad when schools within a district commit to grant programs without the knowledge and consent of the school district's central administration. Often, such action can have negative effects on the school district, because as mentioned earlier, many grantors require some form of contribution to the program by the recipient school. These *in kind contributions* or *recipient match* may commit the district or school to providing resources it doesn't have or cannot afford.

o In kind contributions usually take the form of facilities, materials or personnel assigned to the grant program and paid for by the school. In some cases a recipient match can involve an actual cash contribution to the grant program by the recipient organization. Such cash matches can range from a dollar-for-dollar match to a ratio of one local dollar to ten or more grant dollars.

o Like all programs, grant programs will have operating expenses. It is important for the school district to understand what it will cost to actually run the grant program. Additionally, a school district must determine if it has the infrastructure and operational capacity to run the grant program before it commits to accepting the grant funds.

Responding to an RFP

The grant proposal lays out the essential information upon which the funder will judge the viability of the applicant to complete the program. It is critical for the proposal to convey the information the awarding agency will need to judge the worthiness of the proposed program. Each component or section of the RFP must

Goals	Statement of Need	Procedures	Evaluation	Key Personnel	Budget
What is the aim of the program?	What needs does this program address?	How will the goals be reached?	How will success be measured?	Who will run the program and complete the mission?	What money is needed to implement the program?
Are these goals consistent with the aim of the RFP?	Does this proposal address priority areas per the RFP?	Do the procedures explain how the work will be done?	Is the evaluation designed based on established evaluation standards?	Are key staff qualified to carry out the program?	How will the money be spent? Are proposed expenses allowable?
Are the goals achievable by this grantee?	Will the grant help mitigate the need?	Is the grantee capable of undertaking the procedures?	Does the grantee have the expertise to accomplish the evaluation?	Are personnel assigned to the program for a sufficient amount of time?	Is the budget prudent and reasonable to get the task done?
Are the goals appropriate in scope?	Are the needs within the core duties of the grantee?	Do the procedures align with the needs and goals?	Will the grantee conduct formative and summative evaluations?	Will key personnel be assigned on a full time or part time basis?	What resources will the grantee provide to support the program?
Are the goals long or short term?	Will the needs be resolved or reemerge after the grant expires?	Will the procedures be sustained without grant funding?	Can program outcome be achieved in other settings?	Will the support of key personnel continue after the grant period?	Can the grantee support the program in the future?

Figure 13.1   Common questions in the request for proposal.

be adhered to. Additionally, the specific criteria upon which the proposal will be judged must be addressed in detail.

Some funding agencies will hold pre-application conferences for those considering submitting a proposal. The conference serves to clarify the funder's intent and answer questions about the RFP and grant program. It is a good idea to attend such meetings when they are offered. Here are suggestions to keep in mind when responding to an RFP.

BE CLEAR ABOUT WHAT YOU WILL DO   Unsuccessful grant proposals often confuse the explanation of the organization's need with a description of what the organization will do with the grant (Hall and Hewlett, 2003). Conversely, they often fail to show how what the organization hopes to do will address the needs it has. Grant proposal writers usually do their best to describe how the money will be used. Unfortunately, they often neglect to show how the program they describe relates to the needs they have or the solutions they seek to solve problems. These weaknesses are compounded when the school district operates without a viable strategic plan, as mentioned above.

CHRISTMAS TREES AND MISSION DRIFT   These are common problems among schools and districts that are successful grant proposal writers. A *Christmas tree* school or district is one that pursues any and all grants for which they are eligible without regard to whether the grant program fits into the school's purpose or priorities. Such schools often view grant awards as ornaments, badges of honor or special recognition. These schools fail to judge whether the grant will contribute to the school's goals and fit within its strategic plan. When schools take the approach to apply for all available grants, they often lose sight of their mission or lose commitment to their mission.

Furthermore, when grant programs are layered one on top of the other, the school eventually succumbs to fad chasing. It gives up sound planning and site-based problem solving centered on professional judgment for trendy or flashy programs that function in isolation of each other, with no understanding or context for teachers, students or parents. Some grants can even be counterproductive as they work against established goals. This will eventually become apparent to funding agencies.

UNDERSTAND WHAT DELIVERABLES ARE DUE   Grant programs require awardees to produce a service or product, usually by a predetermined date. These products and evidence of service are sometimes referred to as *deliverables*. Schools that apply for a grant must be clear about what deliverables are due and when they are due. Deliverables represent the commitment that the school district provides in exchange for the grant funds. It is an obligation of the district or school.

ALIGN BUDGET WITH GOALS AND ACTIVITIES   The budget should clearly relate to the activities of the proposal and be justified within the normal operating expenses of the school or district. It is inappropriate and can be illegal to charge items to the grant that are not genuinely related to grant operations. In all cases the grantees must be clear about what is an allowable charge to the grant. The proposal budget is always closely scrutinized by the grantor to make sure all requested funds are reasonable and appropriate, given the proposed goals, objectives and activities of the proposal. The budget should align closely with the activities outlined in the proposal.

INDIRECT COSTS ARE REAL COSTS   School districts are imprudent to overlook the costs associated with receiving and running grant programs. Added expenses such as processing and adding new personnel, telecommunications, copying, receiving and installing equipment, or facilities maintenance are examples of costs that can be overlooked.

Careful planning and negotiating an indirect cost rate with the grantor can help minimize these problems. Here again, the strategic plan helps administrators to keep their eye on the big picture and understand the interrelationships among the various programs and activities of the school or school district.

PROGRAM EVALUATION   This is the systematic investigation of a program's merits (Fink 1995). Most RFPs require an evaluation, but often grantees fail to carry out meaningful evaluations once they get their grant. Frequently, they see the evaluation section of the grant proposal as just another section to complete without regard to how it might actually help the program or their school. In other cases the grant program evaluation is conducted in isolation and thus has little or no connection to the overall effectiveness of the school.

When a school functions with a coherent management style, it uses all available resources for school improvement. Thus, the evaluation requirements of a grant program are used as an opportunity to feed into overall school-wide evaluation and accountability efforts. If a school lacks the expertise to design and conduct sound program evaluations, it should seek outside help from a competent expert. Many grant programs allow this as an acceptable expense.

Proposal Writing Strategies

Administrators, teachers and sometimes parent groups take on the challenge of seeking extra grant funds from various sources in order to meet critical resource needs in their school district, school or classroom. Below is an outline of techniques that have proven successful for experienced grant proposal writers to help them win some extra funding (Ramirez, 1998).

o Grant proposal writing is often a competition, and in order to win the competition, it is important to know basic information about the sponsor of the competition and rules of the game. Be sure you know who the grantor or sponsor of the grant is and what the mission of their organization is. Your proposed program may or may not be compatible with that mission; if it is not, don't bother to apply.

o Novice grant proposal writers often commit several common mistakes that are a waste of time for themselves, their staff and fund-granting organizations. First, determine if you are an eligible recipient. Be sure you are qualified to receive a grant from the organization you apply to before you write the proposal. Grantors are very specific about who they have in mind to receive their grants. Furthermore, you should understand the criteria for funding. Even though your organization is eligible to be a grantee, your proposal idea may not be eligible for funding. Some problems to avoid in this area might be asking for too much or too little money or asking for funds for something the grantor does not fund—for example, asking for construction funds when the grantor has specified grants will go to schools for curriculum development.

o The aphorism "there is no free lunch" applies to grants, so be sure to understand what the deliverables are before you write the proposal. Deliverables are what the grantor expects from you. Some grants can be more trouble than they are worth. The grantee should judge whether the grant

Picture 13.2   Think twice—grant money often has strings attached.

would be a help or burden to their organization before they apply. Grants are not "found money" and always have a price for the recipient.

o   One way to get good at grant proposal writing is to analyze what happened to rejected proposals. In this way, one can learn from failure. Grantors are often encouraging in this regard and interested in helping potential grantees get better at preparing proposals. They often share remarks and rating forms from proposal evaluators and frequently will give advice on how to improve your proposal for the next competition.

o   Get to know the grantor organization and help them get to know your organization or program. If possible, meet with the grantor organization well before the grant competition to learn about their priorities and to share information about what your mission and needs are. Granting agencies like the state department of education and the U.S. Department of Education frequently use outside readers or peers to review proposals. Becoming a proposal evaluator is a great way to become known to the granting agency and learn about successful and unsuccessful proposal ideas. Building a relationship with the grantor organization and grant administrators often pays great dividends.

o   Don't count on need. School people often believe that granting agencies are focused on school needs to the exclusion of proposing a good program that is consistent with the funding agency's mission. It is a mistake to think that need alone will help a school district win a grant. Most grants are not charitable handouts but rather funding for viable educational programs. The grantor often has a goal to develop innovative and replicable programs that can be copied by other schools.

o It is common for a funding agency to enter into negotiations with a school district in order to reshape the proposal and budget. This is the time for the school district to closely consider what it is being asked to do and how much it is being offered to do it. The urge to win the grant competition at all costs is a mistake if the school district finds itself in a position of not being able to fulfill the terms of the agreement in a way that is fiscally prudent and educationally productive.

Successful grant proposal writing need not be a mystery available only to those with secret knowledge. Common sense, a sincere approach and good planning can go a long way to developing winning proposals for grants to help a school or program. Most important is to understand how the grant will help the recipient organization meet its goals and advance improvement.

Managing Grants

When a grant proposal is accepted for funding, the applicant is notified in some official manner. This notification can take the form of a letter or, as in the case of the federal government and some states, a *grant award document*. Foundations and private businesses will provide a grant award letter. The grant award document contains critical information about the amount of money awarded, the length or timeframe of the award, contact persons and account number information needed to draw funds from the bank.

The grant award document represents a financial commitment on the part of the granting agency. At this stage in the proposal and grant process, the applicant is committed to carrying out the program as submitted in the proposal and the granting agency is committed to funding the project. In essence, the two parties have entered into a contract. This is based on the two essential components of any contract—an offer and acceptance of the offer.

Once the school or district is successful in securing a grant, a new set of issues comes into play related to the proper management of the grant. The grant recipient has obligations beyond those specified in the activities section of the proposal. It must also agree to comply with other conditions of receiving the grant money. These obligations are referred to as *assurances* in many government grants.

The granting agency typically has the organization that receives the grant "sign-off" on the assurances. This sign-off usually takes place as part of the documentation included in the proposal. In most cases the chief operating official of the school or district is required to sign. In some cases the awarding agency may ask for official action of the school board as a condition of submitting the proposal.

Sometimes in the jargon of grant writers, this section of the proposal is called "boilerplate." For example, a request for proposal from the federal government will ask applicants to sign off on an array of assurances related to federal civil rights laws, and fiscal and audit requirements. But experienced grant managers understand that boilerplate is serious business given the contractual relationship established between the grantor and grantee. Other terms and conditions can be included or referenced in the letter of notification or grant award document—read it.

Budget and Program Changes

Budget management becomes an important function once the grant is approved. The recipient must determine cash flow needs and establish an accounting process for the grant funds. Most grants restrict the use of the funds and require a clear "audit trail" so the funds received and expended can be readily accounted for. It is common practice for school districts to have a separate fund into which grant monies are received and disbursed. Most federal grants require that grant funds be segregated and not be "commingled" with other revenue sources.

Grantees typically require that any change to budget line items—i.e., individual budget categories—be approved before they take place. A typical example is a request to shift dollar amounts from, for example, the personnel budget line to printing budget line because the personnel line was estimated high in the original grant proposal and the printing line item in the proposal budget was estimated too low. The grantee must understand the rules under which the grant is to be managed in order to know what budget changes can be made at the discretion of the recipient and which changes require prior approval.

Similarly, changes to the educational program originally proposed and approved must also get prior approval. For example, funds for an after-school reading program should not be shifted to support an in-class math tutoring program without the approval of the grantor. Any substantive changes to the activities of a grant program usually require the approval of the granting agency. Be reminded that the needs of the school, no matter how great, are often not of concern to the granting agency. The granting agency's priority is the advancement of its mission through the use of the school. That is why schools cannot use IDEA funding, for example, to patch a leaky roof.

Audits and Reporting

Grant funds are subject to audit. Financial and program audits are routinely conducted of all government grants received by school districts. The school district's annual audit will review the financial integrity of the management of the grant and the implementation of the program in terms of agreed-to activities and legal parameters of the program from an implementation perspective.

The audit will look at expenditures relative to approved activities for the grant program and "allowable expenditures." Therefore, a grant program that restricts personnel costs to classroom teachers would have an "audit exception" if money were used to hire paraprofessionals. In such a case, the school district would be obliged to pay back the improperly used money and may be subject to other action by the grantor. In the case of government grants, this could include criminal prosecution.

Basic data reporting is yet another typical aspect of grants management. Funders are eager to get information about the number and type of participants and other program-related information about the program they are supporting. These data are often used as an indicator of program impact across a state or the nation. In many cases the data are used to justify requests for additional appropriations from the legislature, Congress or a foundation board.

Program evaluation is almost always a part of a grant program. Unlike basic data reporting, program evaluation seeks to determine the impact of the educational program. Evaluation requirements vary from simple reporting of frequency counts to elaborate analysis of student learning impact data gathered from tests of academic achievement. The nature of the evaluation required for each grant is usually specified in the RFP. Properly conducted program evaluations are often complex and costly. The grant recipient is cautioned to keep this in mind when preparing its budget request.

A final consideration in the grant management area is the requirement to disseminate information about the program. Some granting agencies insist that grant recipients actively send out information about the program to like school districts, the media or potential donors of the granting agency. Dissemination obligations may even include the establishment of a demonstration site specifically designed to receive visitors to view the educational program.

A Final Note of Caution

It is advisable for schools and school districts to be selective about which grants to pursue and how many grant programs to undertake at one time. School leaders who view grant programs mainly as "found money" are more likely to run into trouble managing their grants and the overall school program. It is important for the school administrator to keep in mind that grantor agencies have an agenda of their own that they fulfill through the distribution of grant money. Awarding agencies always want something from the recipient. Grants are generally designed as an incentive or motivator for recipient agencies. Grantors want to get the schools to do something on their behalf, e.g., serve a particular kind of student or offer a certain curriculum. School leaders must be able to discern the value of the grant program to the school and judge whether or not to pursue the grant. Contrary to popular wisdom, when it comes to grants, one should always "look a gift horse in the mouth."

Summary

Grant funds can be a helpful source of additional resources for educational purposes. Schools and school districts are well advised to develop skills in seeking and securing grants. But caution is required so that applying for grants is done in a prudent and strategic manner. Grants also have obligations for the recipient and these must be considered before applying. The best approach to building a grant-seeking process is to ensure that grant programs being considered mesh with and support the school or school district's strategic plan.

References

Hall, M., and Howlett, S. (2003). *Getting funded: A complete guide to proposal writing* (4th ed.). Portland, OR: Continuing Education Publications, Portland State University.

Fink, A. (1995). *Evaluation for education and psychology*. Thousand Oaks, CA: Sage Publications.

Ramirez, A. (1998). The winning formula for successful grant writing. *School Administrator's Title I Hotline*, 2–3. Boston: Quinlan Publishing.

Scriven, M., and Roth, J. (1978). Needs assessment: Concept and practice. *New Directions for Program Evaluation, 1*, 1.

Future Trends in School Finance

14

Aim of the Chapter

THE PURPOSE OF THIS CHAPTER IS TO ENCOURAGE the consideration of likely future trends and the direction of policy, politics, litigation and practice in school finance. At the conclusion of the chapter the reader should be able to reflect on the overall content of the text and project his or her own thoughts and opinions about future trends with respect to national, state and local school finance policy and practice. Beyond attempts to predict future direction, the chapter strives to help the reader gain a perspective to craft more effective school finance policies, politics and practices for the future.

Introduction

William Shakespeare reminded us that "past is prologue" (1610/1920), and that perspective is used in this chapter to project the future direction of school funding policy and practice. What should be clear from the readings in this book is that the field of school finance is a dynamic policy environment with a long history of change and development over time. This chapter considers the topics and material covered throughout the book and examines significant trends that will likely affect the direction of policy, politics, practices and school finance litigation in the future.

There are many forces that have a bearing on school finance policy and practice: technology, economics, demographics and the polemics of the American political scene, to name some. Clear trends point to likely destinations that cannot be avoided in the absence of major disruptions to things like politics and economics, or catastrophic events like wars and natural disasters. Futurists in the policy environment use such trend analysis to develop likely scenarios and tendencies in order to anticipate policy direction and policy need, and the possible state of future affairs based on maintaining the status quo or attempting to influence the future (Toffler, 1972; Aburdene, 2007). Some directions in school finance seem inevitable, while others are less clear and dependent on related events like shifts in politics and the economy. Various scenarios are explored in the chapter.

Demographics

The often-used expression that "demographics is destiny" certainly applies in relation to school finance in America. Several issues articulate from this area, not the least of which is the aging U.S. population (Shapner, 2007). The baby boomer generation is retiring at ever-increasing rates and making demands on social programs like Medicare, social security and state retirement systems. Thus, demands on tax revenues to pay for these "entitlement" programs strain budgets and resources that could be available for other programs like school funding.

These obligations are borne by the federal and state government, respectively, so a shift in PK–12 education funding back to the local level seems a likely possibility. This would be a reversal of an almost century-old trend in most states, where funding for schools moved away from a major portion based on local sources to increasing levels of statewide revenues. One impetus for the move away from local revenue sources during the twentieth century was the concern over equity among property-rich and property-poor school districts. It will be interesting to see how far back this trend takes the system.

Federal funding for education has averaged less than 10 percent in most states for more than half a century—this despite the ever-expanding, even overbearing, federal influence on state education policy. Perhaps the federal government will take on a greater share of the cost of PK–12 education, maybe even meeting its obligations under such programs as special education, English language learners, education of the economically disadvantaged and other underfunded mandates. After all, unlike the states, which can only spend the revenue they take in, the federal government can print more money when it needs it.

The emergency funding for education from the federal economic stimulus money of the American Recovery and Reinvestment Act of 2009 (U.S. Department of Education, 2012), authorized by Congress, helped school districts and states shore up depleted state and local revenue sources due to the recession. Many school leaders hope these funding levels will be sustained or even increased. However, the lingering effects of the 2008 economic recession continue to squeeze government budgets at all levels. School funding in the immediate future looks bleak.

Another concern within the demographics realm is the matter of immigrant children in the schools. In an example of history repeating itself, schools in the twenty-first century are seeing an influx of foreign-born children on a par with that seen at the end of the nineteenth century. Once again, America will ask the public schools to shoulder the major portion of the task of educating these new Americans.

From a national perspective, immigrants are integral to the economic viability of the country and a part of the solution to the growing entitlements due the aging population. The declining birth rate among the native population only serves to underscore the inevitable growth of numbers among immigrants to the United States. The question arises when considering these developments: Will the funding be there to meet the challenge?

School Choice and Privatization

It is difficult to predict whether the trend toward more school choice will be sustained. Some forms of choice, like charter schools and tax credits for private school tuition, are popular with many policy makers on both ends of the political spectrum. From that perspective, it would seem that predicting an expansion of choice options for the future would be a safe bet. However, evaluation studies of charter schools and voucher programs tend to show no or minimal student achievement gains (Center for Research on Education Outcomes, 2009; Ladd, 2003). Inevitably, questions about the efficiency of the proliferation of choice options will grow more frequent and louder.

However, the American ethos, rooted in liberty and freedom, will likely make it difficult to reverse this trend. Privatization advocates have seen uneven progress in this aspect of choice as policy makers have enthusiastically rushed in to this realm of choice, only to beat a hasty retreat as questions about the propriety of using tax dollars to support private corporations and profit making from the education of children.

If the private sector can deliver on its claims to produce higher student learning at a lower per student cost, it will no doubt win the day. The pressure to reduce or at least slow down the rate of growth in the cost of education grows each year; thus, less expensive choice options become more attractive to political leaders as the pressure builds to find more funding for public education. However, time will tell if the choice experiments in the private sector will be sustained and taken to scale.

Technology

The popularity of online education combined with its cost effectiveness relative to face-to-face education programs augurs well for the expansion of online education and the use of technology (Wiesenberg and Stacey, 2005; Parsad and Lewis, 2008). This rapidly growing field of education will likely expand even further as technology improves—making distance education even more convenient for student and teacher—and becomes more personalized (Ramirez, Burnett, Meagher, McMullen-Garcia and Lewis, 2009).

The generational divide between the new "connected" youth and the "old-school" types will disappear as technology further permeates society. Technology also has the potential to diminish issues of equity among school districts and schools with respect to curriculum, services and programs. Data sharing and improved methods of data transmission may offer some improved efficiencies for the PK–12 system, but this seems marginal in consideration of all the types of data manipulated in the public school environment. The exception here might be online student testing, once initial start-up costs are made up.

Special Populations

Renewed attention will turn to students with disabilities, English language learners and economically disadvantaged students as it becomes ever more apparent that economic investment in such populations pays great dividends. We can also expect that litigation with regard to the adequacy of funding for such groups will increase and achieve some success in light of mounting evidence from production cost function studies (Patrinos and Psacharopoulos, 2002). Such studies continue to show that the investment in human capital made by a nation is thoroughly rewarded with increased productivity in the economy, improved global competitiveness and improved social conditions in many aspects of society (Psacharopoulos, 2006).

Perhaps one population that may see an increase in resources is students with gifts and talents. This population has received scant attention after the Cold War education policies of the National Defense Education Act of the late 1950s. However, for similar reasons in different times, the nation's attention may once again turn to students with exceptional gifts and talents.

As the hugely populated developing nations of China and India expand their education systems, the sheer numbers of talented individuals who will be able to take advantage of new educational opportunities will be enormous. Among those masses of new students will inevitably be individuals with extraordinary capabilities who will contribute enormously to their respective nations. With these countries having populations four times that of the United States, it is expected that they will be able to develop four times as much exceptional talent. This realization will renew the brain race in America.

Globalization

The planet continues to shrink at an ever faster pace as technology and economics combine to make human capital ever more transportable. The consequence of increased globalization means ever higher demands on the output of the American education system. The demand for an educated citizenry and globally competitive workforce will translate into more spending for education at every level (Checchi, 2006). Since the 1990s, there has been a push for standards in all aspects of industry and commerce, which has slowly made its way into the field of education.

Most professionals and policy makers in the PK–12 work environment would dispute the idea that standards have been slowly adopted in education. For some time now, every state has adopted some form of academic content and performance standards and elaborate assessment systems to measure student progress relative to such standards. Unfortunately, these standards are typically not benchmarked to international levels, much less to the performance of the academically top-producing countries in the world. Furthermore, the standards established thus far are limited to academic content for the most part and do not address the broad system of education.

Globalization, which is unavoidable like the march of time, will require the United States to make systemic changes in the structure of its PK–12 system of education. The old-fashioned, quaint notions of the 180-day instructional year; the incoherent patchwork curriculum; and the weaknesses of admission to teaching, pre-service and in-service education, which allow critics to call teaching a specious profession, will all be wiped away.

As was the case over 100 years ago when the education systems in the various states began to revise the standards of the day and address these very same issues, the nation will now finally be forced to make the systemic changes needed in PK–12 education and meet the standards of a twenty-first-century education system for a leading developed nation. The idea that tinkering with the nineteenth-century model U.S. education system will render twenty-first-century results is coming to an end.

A simple analogy is a comparison to the United States Olympic teams. When the country lost its competitive edge in many sports in the 1950s and 1960s, it made a decision to move away from the old way of selecting and training its athletes. The idea that sufficient preparation could take place a few months before the games or that "amateur" athletes could be held to a different, traditional standard in the United States, as compared to the international standard, led to disappointing results in Olympic competition in many sports. The restructured U.S. Olympic Committee's selection and training approach has seen the country reassert its dominance in these athletic endeavors. The PK–12 education system will do the same.

Teacher Recruitment and Compensation

The biggest and most expensive future change facing the nation within the PK–12 education system is the restructuring of the teaching profession. As mentioned several times already in this chapter, major systemic changes must take place, but such changes will have little effect without a significant change in the trend lines related to who is recruited to be a teacher, who chooses to be a teacher, who gets to be a teacher and the career path of those who do teach. The educational demands of the twenty-first century require that intelligent, well-educated and highly motivated individuals be attracted to teaching. And finally, pre-collegiate teaching must be reshaped into a full-fledged profession like law, medicine or engineering.

Once again, the nation will need to abandon its nineteenth-century model of the teaching profession. The teaching profession must become economically competitive with other professions that require postgraduate training. In order to attract capable, talented individuals from across the talent pool, changes must occur. At a cultural level, Americans must change their perspective about the status of the teaching profession. The nations that beat the United States on international edu-

Picture 14.1   Public education must attract and retain intelligent, educated, talented and committed people to teaching.

cational assessments have several things in their favor; among them is that teaching is a respected and highly regarded profession in those countries. Competition for entrance into teacher education programs is stiff, training is rigorous and a teaching position is not easily earned. Compensation in these countries is on a par with other professions that require similar education and training.

As discussed in chapter 9, teacher salaries account for almost one-half of school district operating budgets. The idea of making teaching economically competitive with other professions is a daunting thought. But no amount of technology or standards or assessments will make a difference in the educational outcomes of the nation unless the teaching force is greatly improved. For those currently in teaching, and those entering in the future, it means large-scale, ongoing and substantive in-service education. For those who strive to enter the profession, it means entering a competitive process for what will be highly desirable teaching jobs.

The cost of making this transition will be enormous. In round figures, the price tag shapes up like this: $600 billion for annual operating expenses for PK–12 education; $300 billion for salaries for instructional personnel; a 30 percent increase in salaries to start ramping up to a par level with other professions; net cost will be about $100 billion additional to the annual cost of the PK–12 system. Creating a competitive market for the teaching profession is a pivotal future challenge for the nation. Continuing with the same system and the tinkering disguised as education reform will not accomplish the kind of shift necessary to get the results demanded of twenty-first-century schools.

Centralization or Localism

The expanding reach of the federal government in the PK–12 arena has been a steady trend since the 1960s. The detailed programmatic aspect of this involvement culminated with the reauthorization of the Elementary and Secondary Education Act in 2001 under the new name the No Child Left Behind Act. Currently due for reauthorization again, it remains to be seen whether the intrusive parts of the bill will be scaled back or enhanced.

The history of federal education efforts at the pre-collegiate level have been numerous and varied. Clearly, federal policy makers have historically been supportive of the nation promoting and expanding education, but without direct administration by the national government. Exceptions to the direct operation of schools by the U.S. government have occurred at different times, for example, the military academies during the early nation-building period, the Freedmen's Bureau during the post-Civil War era, the Bureau of Indian Affairs schools established after the Indian wars of the late nineteenth century and the Department of Defense Dependent Schools for dependents of military personnel and oversees foreign service workers. However, for the most part the neighborhood school serving the average American family has been the domain of state and local government.

This is not to say that local schools are untouched by federal education policies or programs—to the contrary. Most public schools get some form of federal assistance, be it school lunch subsidies, ESEA Title I funding or grant funds for some other special population. However, these funds have tended to be limited, targeted and often accepted by school districts on a voluntary basis. Exceptions to this approach have been in the area of civil rights legislation, which have directed facets of school policy and operation in the interest of some protected class, e.g., students and employees with disabilities.

Today we see advocates promoting national education standards and a national system of assessment for all public schools (Isaacson, 2009). Critics of this trend see it as federalizing the nation's pre-collegiate education system and being a fast track to disaster (McCluskey, 2009). The idea of the president as national superintendent of schools and Congress as the nation's school board sends shudders among some in the country. However, the popularity of this idea among the American public is strong (Bushaw and Gallup, 2008) and so is the political momentum.

Implications from the school finance perspectives could see interstate equity litigation in which plaintiffs will argue for equal funding comparable to neighboring states in the region or some national average. Adequacy cases could also see a resurgence as poor states seek funding on a level with states that offer more extensive programs and educational service. The question remains in such a scenario: will the federal government fill the financial void, or will states see even more unfunded mandates from Washington?

Adequacy and Equity

The past several decades have seen many advances in the conceptualization of theories about how equity and adequacy in school finance have been defined. Perhaps none of these advances has been more significant than the idea of linking student educational outcomes to funding. Through such analyses as successful school comparisons, expert panel school development and statistical methods like production functions and stochastic frontier modeling, school finance experts have been able to demonstrate deficiencies in equity and adequacy in current school finance systems. Further developments with these methodologies should be expected.

Streaming concurrently with this trend is the further development of theories associated with systemic thinking about the relationship between standards, assessment systems and adequate funding. Building on the work of Marshall S. Smith and Jennifer O'Day (1990), researchers from numerous academic fields have been able to corroborate the theories about systemic school reform and improved study learning in a context of equal education opportunity for all students. States like Kentucky, Massachusetts and Wyoming have undertaken trailblazing work in this area. More recent efforts in New York and New Jersey underscore the complexity and sweeping range of such policy and funding systemic reform.

Within the considerations of adequacy, the issue of quality preschools for all who want them must finally be addressed. Current public education preschools overwhelmingly target at-risk and disabled students. As a nation, we cannot expect to compete internationally when so many of our children lag behind because of inadequate education options during such a critical developmental stage. As described earlier, our demographics have shifted tremendously, but we have not made the policy changes to match our circumstances. We do not live in the 1950s era when mom stayed home and nurtured the children, while dad went to work and earned a wage that sustained a middle-class existence. Economic conditions today require two incomes for most families to maintain a middle-class existence.

Today, preschool and child-care options are too often of low quality, except for the lucky few who get into good publicly supported programs or the well-off who can afford the expensive ones. National policy addressed this issue almost two decades ago when Goals 2000 (1994) was adopted by Congress and signed by the president. Many states followed suit by adopting the national goals and even enshrining them in state statute. One of the goals declared that, "All children would be ready to learn." It is estimated that to achieve this goal, funding would have to increase at all levels—state, local and federal—by 30 percent, or another $200 billion. Most policy makers rarely talk about universal preschool, much less Goals 2000.

Picture 14.2   The courts will continue to play an important role in school finance.

The equity/adequacy trend proffers even more litigation efforts among lagging states with respect to seeking an equilibrium among systemic reform, equal educational opportunity, equity in school funding and adequacy of resources to meet newly established benchmarks of a twenty-first-century education. The seemingly slow iterative process that moves from state to state will possibly be accelerated by federal education policy or breakthrough litigation. For example, a modern-day watershed case, on a scale of the *Serrano* cases in the 1970s, could set off a chain of litigation or legislative actions that would leapfrog the PK–12 system along the systemic reform trend line. Some scholars argue, forcefully, that the time is ripe to revisit a *Rodriguez*-type case to establish legal standards of equity at the national level (Saleh, 2011).

Revenue for Schools

The one big question about future trends in school finance is the proverbial elephant in the room: How will we pay for all this? History tells us that whenever we see a large increase in revenue for schools, we also see a backlash in the form of taxpayer revolts and attendant tax policies. The *Serrano* case in California influenced politicians in other states to address inequities in their state funding frameworks. However, according to some policy analysts, this also led to state level tax revolts like Proposition 13 in California; Massachusetts saw its anti-tax coalitions form around similar circumstances, and Colorado added an amendment to its constitution through a voter initiative called the Taxpayer Bill of Rights, all of which were directed at constraining revenue growth, much of which was slated for school funding.

As demographic shifts in the population, competing demands for government service, the effects of globalization and restructuring in the economy take place, new ways of funding education will emerge. Likely new revenue streams will be developed through such mechanisms as excise taxes on the "new economy," i.e., newly developed energy sources and new technologies. A federal sales tax, which has been discussed for years, seems inevitable. And one should not discount some of the old tried-and-true ways for government to collect funding, through gambling and other "sin" taxes. The states have led the way here with the proliferation of lottery and casino gambling. We should not be surprised to see the federal government take its cut, perhaps in the currently wide-open field of online gaming and sports betting.

What was unthinkable a generation ago seems poised to become the norm. Medical marijuana, now widely available in states like California and Colorado,

Picture 14.3   Government will continue to bet on "sin taxes" for expanded revenue sources.

may be the proverbial "camel's nose under the tent." At the time of this writing, states with such legal provisions are struggling to develop systems to regulate the growth, distribution and sale of cannabis. In another example of what could be history repeating itself, whispers about the wisdom of prohibition against marijuana use are being heard in cash-strapped legislatures around the country. As with the Volstead Act of 1920, which prohibited alcohol sale and consumption, policy makers are contemplating the idea of reversing laws against marijuana that cost all levels of government billions of dollars a year to enforce. Might we see local marijuana stores, like liquor stores, in neighborhoods across the country, generating billions of dollars in new taxes? Colorado is well on its way.

The point here is that the local, state and federal government will do well to hang on to existing revenue sources and will have to find supplements to current sources of residential property tax, personal income tax, corporate income tax and sales tax. The demands for government services like health care, national defense and infrastructure are increasing with no end in sight. The PK–12 education system is in line with the rest with its hand out, too—and it is a big hand. New sources of revenue for governments at all levels will be established.

Summary

A good predictor of future trends is past trends, and the area of education finance is no exception to this concept. Any number of existing phenomena shape events in ways that gather momentum and then organize toward new directions. Some of these phenomena help to predict the future. Demographics, advances in technology, politics and globalization are examples of trend-shaping occurrences.

The shifting profile of the American population will once again affect events in the public schools. Whether it is the changing nature of the student population or the aging of the population that provides the tax revenue to support the schools, the twenty-first century will not be the same as the twentieth in this regard. Declining birth rates, childless households, an advancing median age for the nation and increased immigration will all have a bearing on the question of continued support for public education.

Technology continues to offer the promise of solutions to the challenges of improved student learning and more efficient operation of the PK–12 education system. However, after more than three decades of accelerating connectivity and ever more sophisticated hardware and software, technology still remains a promise. Expectations remain high that technology will deliver on its promises.

School choice and privatization have had popular and political support over the past several decades, yet results from these policy "experiments" are unclear. Several trends in this area suggest a forthcoming decision point about educational choice for policy makers in the near future. The first of these trends is that over the past twenty years, no state that has had a referendum on universal school vouchers has approved such a system. In fact, when faced with this question, voters have overwhelmingly turned down such policy proposals. Therefore, legislative action has been the driving policy force for this form of choice.

The second trend is the growing number of students being educated under some kind of choice policy program. The choice movement is fast approaching a critical mass in the number of students so educated, and at that point, the experimental nature of these programs will lose the cover of "let's see if it works." Soon it will be time to evaluate the viability of choice policy based on results.

Teaching is at the heart of the educational process, and it seems as though every generation must rediscover this fact. The American educational system will not improve beyond the incremental change it has seen over the past fifty years unless the teaching force across the nation is substantially altered. In order to attract an intelligent, well-educated and motivated teacher corps, two things must change: the attitude of the American people toward teaching as a profession must shift to one of high regard, and this must become conspicuous in the compensation system. In other words, to attract and retain top-level professionals to education, the status and compensation for teachers must be restructured.

Is the United States of America a democratic federal republic or not? This question will dominate the discourse about American education policy in the near future. In the late nineteenth and early twentieth century, education reformers focused on driving politics out of the education system through structures like nonpartisan governing officials and the appointment of professional managers. Their goal was to drive out the forces of political patronage, cronyism and corruption and put education on a professionally run basis. We have lost sight of those policy objectives.

The trend toward centralization of education policy in Washington, D.C., speaks to an arrogance of power that is manifested within the two-, four- and eight-year election cycles for Congress and the executive branch. The iterative policy creep from America 2000, to Goals 2000, to No Child Left Behind shows a clear and disturbing trend for those who believe all policy wisdom for the American education system does not reside within the Washington, D.C., beltway.

There is a role for the federal government in education, but it is not in pulling the strings of state education agencies, dictating curriculum or testing all American students. The federal government has a duty to ensure that the full force of the U.S. Constitution protects the rights of parents, students and educators.

Individual states are capable of designing "world-class" education systems and preparing students with "twenty-first-century" skills. Those politicians in states that choose to offer their citizens a second-class education will live with the consequences. One of the strengths of our education system has been the proliferation of ideas, innovations and programs that emerge from the school districts and states across the country. The whole nation risks decline under a centralized system if Washington gets it wrong. We should expect to see a vigorous dialogue about the centralization of education policy in the near future.

Just as it has been since the colonial period in America, the question of funding for education will be an ongoing issue. How much, for whom, who pays and for what will continue to dominate the discussion. Despite great advances over the past one hundred years in the policy arena for equity and adequacy concerns, the final word has not be spoken. Leadership from the executive, legislative and judicial branches of government at the local, state and federal levels will contribute to the contours of new policies. When considering all that has been written in the above paragraphs, what should be clear is that sufficient funding for education will remain an ongoing question and the distribution of those funds an ongoing debate.

References

Aburdene, P. (2007). *Megatrends 2010.* Boulder, CO: Sounds True.

Bushaw, W. J., and Gallup, A. M. (2008). Americans speak out: Are educators and policy makers listening?—The 40th annual Phi Delta Kappa/Gallup poll of the public's attitudes toward the public schools. *Phi Delta Kappan, 90*(1), 9–20.

Center for Research on Education Outcomes (2009). *Multiple Choice: Charter School Performance in 16 States.* Stanford, CA: Center for Research on Education Outcomes, Stanford University.

Checchi, D. (2006). *The economics of education: Human capital, family background and inequality.* Cambridge, UK: Cambridge University Press.

Goals 2000 (1994). H.R. 1804, Goals 2000: Educate America Act. Retrieved from http://www2.ed.gov/legislation/GOALS2000/TheAct/index.html

Isaacson, W. (2009, April 15). How to raise the standard in America's schools. *Time.* Retrieved from www.time/magazine/article/0,1971,1891741,00.html.

Ladd, H. F. (2003). Comment on Caroline M. Hoxby: School choice and school competition: Evidence from the United States. *Swedish Economic Policy Review, 10,* 67–76.

McCluskey, N. (2009, March 18). Let's not play standards roulette. *Cleveland Plain Dealer.* Retrieved from http://www.cato.org/pub_display.php?pub_id=10055.

Parsad, B., and Lewis, L. (2008). *Distance education at degree-granting postsecondary institutions: 2006–07* (NCES 2009–044). Washington, DC: National Center for Education Statistics, Institute of Education Sciences, U.S. Department of Education.

Patrinos, H. A., and Psacharopoulos, G. (2002). Returns to investments in education: A further update. *World Bank Policy Research Working Paper No. 2881.* Retrieved from http://ssrn.com/abstract=1273483.

Psacharopoulos, G. (2006). The value of investment in education: Theory, evidence, and policy. *Journal of Education Finance, 32*(2), 113–136.

Ramirez, A., Burnett, B., Meagher., S. S., McMullen-Garcia, J., and Lewis, R. (2009). Preparing future school leaders: How can it be accomplished online? *National Council of Professors of School Administration Yearbook, 2009.*

Saleh, M. (2011). Modernizing San Antonio Independent School District v. Rodriguez: How evolving Supreme Court jurisprudence changes the face of education finance litigation. *Journal of Education Finance, 37*(2), 99–129.

Serrano v. Priest, 5 Cal. 3d 584, 96 Cal. Rptr. 601, 487 P.2d 1241. (1971).

Shakespeare, W. (1610/1920). *The tempest.* Cambridge, MA: The Harvard Classics.

Shapner, S. L. (2007). *Demographics of the United States.* New York, NY: Nova Science.

Smith, M. S., and O'Day, J. (1990). Systemic school reform. *Journal of Education Policy, 5*(5), 233–67.

Tallent-Runnels, M. K., Thomas, J. A., Lan, W. Y., Cooper, S., Ahern, T. C., Shaw, S. M., and Liu, X. (2006). Teaching courses online: A review of the research. *Review of Educational Research, 76*(1), 93–135.

Toffler, A. (1972). *The futurist.* New York, NY: Random House.

U.S. Department of Education (2012). American Recovery and Reinvestment Act. Retrieved from http://www.ed.gov/policy/gen/leg/recovery/index.html.

Wiesenberg, F., and Stacey, E. (2005). Reflections on teaching and learning online: Quality program design, delivery and support issues from a cross-global perspective. *Distance Education, 25*, 385–404.

Index

About the Author

Al Ramirez is professor in the department of Leadership, Research, and Foundations at the University of Colorado, Colorado Springs. His transition to university-level teaching follows a twenty-five year career in PK–14 education. His experience includes positions as a teacher, counselor, principal, central office administrator and superintendent of schools. He has also held key education policy positions in the Nevada and Illinois state departments of education and served as chief state school officer in Iowa.

Dr. Ramirez has published widely on a variety of topics in such journals *as Phi Delta Kappan, Executive Educator* and *Educational Leadership* and has presented at numerous state, national and international conferences. Dr. Ramirez also served appointments to several national education advisory boards and commissions. His consulting work in policy and education finance has a client list that includes foundations, governments, school reform organizations and education entities.